THE ENGINEERING PROFESSION

Its Heritage and Its Emerging Public Purpose

Dan H. Pletta, P.E.

UNIVERSITY
PRESS OF
AMERICA

LANHAM • NEW YORK • LONDON

Copyright © 1984 by

University Press of America,™ Inc.

4720 Boston Way
Lanham. MD 20706

3 Henrietta Street
London WC2E 8LU England

Printed in the United States of America
ISBN (Perfect): 0-8191-3836-3
ISBN   (Cloth): 0-8191-3835-5

All University Press of America books are produced on acid-free
paper which exceeds the minimum standards set by the National
Historical Publications and Records Commission.

Table of Contents

Preface

All books have a purpose, at least in the eyes of
their authors. I wrote this book with the objective of
providing engineering students and practitioners, as
well as the general public, with a more precise de-
scription of the engineering profession, with a brief
treatment of its heritage, and with an explanation of
its emerging public purpose. You may wonder why I felt
that was needed. There were three reasons.

First, I sensed that many audiences had a hazy understanding of the topics discussed in this book. I
have presented these ideas and opinions at engineering
society meetings, at student seminars, and at faculty
meetings over the past several decades throughout the
United States and a few foreign countries. Frequently
definitions of such concepts as professionalism, technician, capitalism, and even of engineering were too
varied for coherent discussion. These differences had
first to be resolved.

Second, I encountered little awareness of the need
for engineers to get involved in societal leadership,
as distinct from either administrative or engineering
management. Civilization today is so dependent upon
technology that it cannot afford to ignore the advice
engineers could contribute when societal missions like
civil defense, pollution control, nuclear power, and
even resource depletion are formulated. Engineers
should not only be aware of the need for their leadership, but they should also recognize their duty as professionals to inform the public about the benefits,
limitations, and side-effects of technology. If engineers will not lead, they should not complain about being led, even if those who then shoulder that responsibility are less prepared to understand how fragile our
technological civilization is. As society grows more
complex, it approaches ever closer to critical instability.

Third, I finally realized that much of the lack of understanding could be traced to a lack of instruction in professionalism for engineering students. Having been engaged in the teaching of engineering for almost half a century I must assume part of the blame. But four year curricula cannot be all things to all people. Four year programs might have sufficed fifty years ago but cannot hope to educate engineers today for practice, management, and leadership spanning the next fifty years. To realize that objective, life-long continued education will need to be restructured by initiating intimate industrial-collegiate cooperation.

There did appear to me to be a need for a brief discussion of many subjects so that engineering professionals could grasp the scope of their heritage and of their obligation to society. This book is an attempt to present these topics as coherently as possible and to consolidate the pertinent discussions of them taken from some of my own previously published papers. In some instances I have quoted rather freely from them so as to give adequate recognition to the journal in which they were first published, and to save the reader's time when a simple reference alone would be inadequate to convey the meaning of the topic.

The book also provides those readers who may wish to explore other references in depth with a rather extensive bibliography which I have cited extensively. It was from these references and from my contacts with other engineers, scientists, scholars, friends and acquaintances, and my own professors, that I gathered the ideas expressed here. I owe all of them a debt of gratitude, and can only hope that I have acknowledged their help sufficiently well.

There are, in addition, five people at Virginia Tech, from whose engineering faculty I retired over a decade ago, who deserve special acknowledgement. George A. Gray, its associate dean of engineering, and William W. Payne, an emeritus professor of civil engineering, were particularly helpful in discussing the topics, reviewing sections of the manuscript, and in suggesting changes. Secretaries Nancy Linkous, Charlene Christie and Connie Callison were indispensable for typing and producing both the original manuscript and the final version of the book on a word processor. In the final analysis, I assume responsibility for the contents and for any omissions or errors where they may occur.

Perhaps a word about some of the statements and of the philosophy of the contents is in order. During my lifetime I have watched how superbly technology has flourished in our free, democratic society. It nurtured a capitalistic economic system which enabled the United States to become the most powerful, richest, and most generous nation in history.

There is no guarantee, however, that such a political environment will endure. Economic, political, and scientific forces are perpetually producing an ever more complex and potentially unstable technological civilization. Ponder, for a moment, the chaos that would ensue if electric power were suddenly decreased significantly or cut off completely. It is in the public arena where those who understand technology, and who base their decisions on logic rather than emotion must be groomed and encouraged to participate in a cooperative leadership role. It would do little good for industrial mechanization to shorten the work week and to increase output if workers are not taught to appreciate their leisure, and if the surplus goods produced are not distributed equitably.

Past technological and political trends have made me both optimistic about the future, and yet quite concerned. There can be real hope if those who understand technology will shed their complacency about the need for leadership, about the increase in governmental regulations, and about the gradual loss of freedom to practice as a responsible professional.

The statements I have made, or quoted, in the book are there to raise questions, to encourage discussion, and to present my ideas of possible solutions. They are not meant to be singular solutions or dictatorial edicts. Whether enough discussion will result in the improvement of education, and unite the practitioners into a professional force better able to serve society, only the future can tell. The opportunity exists now to make it so.

I found that students particularly were vitally interested in the professional concepts expressed in these ideas. Being young, they could still afford the luxury of idealism. But it is this idealism which the world must learn to blend into practical societal goals if civilization is to continue its progress.

Society today could benefit from a renaissance of engineering leadership that would prevent unnecessary side-effects of technology while continuing its amazing benefits and improving its distribution. Such efforts will involve politics and economics as well as engineering practice at ever advancing scientific levels. Engineers must be involved in the formulation of these highly technical missions. They cannot be excluded or remain too passive without jeopardizing their own safety and society's freedom. Neither can they exert much political influence unless they act rather than react, and unless they unite professionally.

If this book serves to inform the public, to educate students, and to motivate engineers to strive for professional behavior that is expert and ethical enough to enlist the public's trust, it will have served its purpose.

Dan H. Pletta, P.E.
Blacksburg, VA
October 1983

Tables and Figures

Acronyms and Abbreviations

AAAS	American Association for the Advancement of Science
AAE	American Association of Engineers
AAEE	American Academy of Environmental Engineers
AAES	American Association of Engineering Societies
AASHTO	American Association of State Highway and Transportation Officials
ABA	American Bar Association
ABET	Accreditation Board for Engineering and Technology
ACE	Association for Cooperation of Engineers
ACI	American Concrete Institute
ACS	American Ceramic Society
AEAA	Association of Engineers and.Architects of Austria
AEC	American Engineering Council
AES	American Engineering Society
AIA	American Institute of Architects
AICE	American Institute of Consulting Engineers
AIChE	American Institute of Chemical Engineers
AIEE	American Institute of Electrical Engineers, see IEEE
AIME	American Institute of Mining Engineers
AMA	American Management Association
AMA	American Medical Association
APWA	American Public Works Association
AREA	American Railway Engineering Association
ASAE	American Society of Agricultural Engineers
ASCE	American Society of Civil Engineers
ASEE	American Society for Engineering Education
ASEM	American Society of Engineering Management
ASHRAE	American Society for Heating, Refrigerating, and Air-Conditioning Engineers
ASHVE	American Society for Heating and Ventilating Engineers (now ASHRAE)
ASME	American Society of Mechanical Engineers
ASNE	American Society of Naval Architects
ASTM	American Society for Testing Materials
BAEC	British-American Engineering Conference
BART	Bay Area Rapid Transit (San Francisco)
BM	Bureau of Mines

BTU	British Thermal Unit
CAB	Civil Aeronautics Board
CEC	Cleveland Engineers Club
CES	Council of Engineering Societies
CPA	Consumer Protection Act
CPA	Certified Public Accountant
CPC	Committee on Professional Conduct - ASCE
CPSC	Consumer Product Safety Commission
DDS	Doctor of Dental Surgery
DED	Department of Education
DOJ	Department of Justice
DVM	Doctor of Veterinary Medicine
EC	Engineers Council
ECPD	Engineers Council for Professional Development
EIT	Engineer-in-Training
EJC	Engineers Joint Council
EP	Engineering Professional
EPA	Environmental Protection Agency
EUSEC	Conference of Engineering Societies of Western Europe and the U.S.A.
FAES	Federated American Engineering Societies
FDA	Food and Drug Administration
FHA	Federal Highway Administration
FMC	Federal Maritime Commission
GE	General Electric Company
GEC	General Engineering Conference
GNP	Gross National Product
HEW	Department of Health, Education and Welfare
HHS	Department of Health and Human Services
IAI	International Alliance of Ingeniors
IBM	International Business Machines, Inc.
ICC	Interstate Commerce Commission
ICE	Institute of Civil Engineers (Great Britain)
IEC	International Engineering Congress
IEEE	Institute of Electrical and Electronic Engineers
IQ	Intelligence Quotient
IRE	Institute of Radio Engineers, see IEEE
JCC	Joint Conference Committee
JD	Juris Doctor
MBA	Master of Business Administration Degree
MD	Medical Doctor
NAE	National Academy of Engineers
NCEE	National Council of Engineering Examiners
NSF	National Science Foundation
NSPE	National Society of Professional Engineers
OSHA	Occupation, Safety and Health Administration
OTA	Office of Technology Assessment, U. S. Congress
PE	Professional Engineer
R&D	Research and Development

RN	Registered Nurse
ROTC	Reserve Officers Training Corps
TNE	Transnational Enterprises
UES	United Engineering Societies
UID	United Ingenieur Diplomates
UPADI	Union Pan Americana des Associaciones des Ingenieros
WEC	World Engineering Conference
WFED	World Federation of Engineering Organizations

The Purpose and [a]
Obligations of Professions

1A. THE PURPOSE

Whether engineering is a well defined profession and has a definitive public purpose is debatable. Many engineers merit professional status; many probably do not. The entire mass of those who claim the title of engineer has not yet congealed. Some who usurp the title are unqualified. All learned professions do have a public purpose, or society would not consider them as such. This does not mean that all individuals who call themselves professionals--say mechanics or athletes-- have a profession to which they can belong, learned or otherwise.

Like all vocations the engineering profession does have a heritage. Now it also has a challenging obligation to accept an enhanced leadership role and help formulate society's technological missions. The public purpose of the engineering profession will, therefore, be assumed here to be one:

a)   that will serve the public by using the materials and forces of nature for man's comfort and benefit;

b)   that will provide a corps of professional practitioners eminently competent technically, whose foremost dedication is that of unselfish service to society; and

c)   that will develop a nucleus of engineers to complement other professionals for the

---

aPresented as part of the opening lecture of a special course on "The Engineering Profession" at the Illinois Institute of Technology, Chicago, January 26, 1977, under the title of "The Engineering Profession and Its Public Purpose."

industrial, governmental and societal leadership of our technological civilization.

The public purpose of engineering was not always that broad. It has been changing slowly, perhaps almost imperceptibly, during the past century. Previously, when populations were less crowded and capital invested in engineering projects was smaller, designers could afford to take greater risks. Their role then was to serve, rather than to lead society. They produced those goods and services society wanted and could afford, and concerned themselves mainly with the major aspects of public health and safety, particularly on major public projects. Florman (4) tells how they enjoyed that role, and hoped that their effort would free man from want, enhance his concern for his fellowmen, and improve society. A brief history of this engineering effort, and of engineering education, is discussed more fully in Chapters 3 and 6.

There was less need formerly to worry about disturbing the environment or conserving resources. There were plenty of open spaces and more than ample resources. But our wants then were more modest, our know-how more primitive, our productive capacity far smaller, and our population growth less explosive. There was enough produced then worldwide to go around, although its distribution has always been skewed so that 5% of the people earn and consume 30% of the production, while 30% live in want and starvation.

As society became more affluent, it became increasingly concerned about the impact of technology on the environment, and of any ill effects consumer products and services might have on its health and safety. Recently, a few members of society have also become alarmed about the world's dwindling natural resources, but most people are still unconcerned.

Nevertheless, the public purpose of engineering has broadened as engineering has become more professional; i.e., more conscious of the need to protect not only the public's health and safety at all times, but also to preserve its environment and to conserve its material resources.

J. Douglas Brown, an economist by profession and former Dean of the Faculty at Princeton University, emphasized this need for change while speaking at the dedication ceremonies of its new Engineering Quadrangle in 1962. He counseled engineers then to "throw aside

2

the last vestiges of the engineering profession's evolution from a craft and take on the full responsibility
of a learned profession." He felt that "The professional engineer, more than anyone else, must act as the
moderator and the interpreter between the two worlds of
science and the humanities" (11).

Not all engineers are professional or independent
enough to heed that calling. The employment status for
engineers has changed mainly from that of an independent consultant to that of a corporate employee during
the past century. A few employees have succumbed to
pressures from a few shortsighted managements to over-
look any questionable effects that their designs might
have on the public's health or safety--or environment--
in favor of corporate profits and their own jobs. Some
who did not succomb were fired. A few were helped by
their professional society.

Professional mechanisms, like codes of ethics or
the counsel of ombudsmen, that might have corrected
these malfunctions individually at their source, were
either poorly enforced or non-existent. Society then
had to resort to collectivized legal action to remedy
the situations by creating watch-dog bureaucracies that
now constrain all of industry. A few, like FDA, BM,
CAB, OSHA, CPSC, are listed in Table 1A. Their cost to
society will be treated in detail in Articles 5B and E.

TABLE 1A - A FEW GOVERNMENTAL AGENCIES
CONCERNED WITH HEALTH AND SAFETY

1906	Food & Drug Administration	FDA
1910	Bureau of Mines	BM
1945	Civil Aeronautics Board	CAB
1953	Department of Health, Education & Welfare	HEW*
1970	Occupation Safety & Health Administration	OSHA
1970	Environmental Protection Agency	EPA
1972	Consumer Product Safety Commission	CPSC
1979	Department of Education	DED
1979	Department of Health & Human Services	HHS
	Consumer Protection Act**	CPA

*Abolished 1979 **Unauthorized to date.

This drift from individualized to collectivized
action by professions and by governments now restrains
engineering practitioners and industries alike. The
more constraints that are placed on industry by governmental bureaucracies, the less freedom business and the
public have. As constraints continue to grow, business
ultimately loses its competitive advantage in world

3

trade; and people see their governments transformed into fascist states, if people retain ownership of industry; or into socialist states, if ownership is nationalized. The form of such governments may remain democratic or change to a bureaucratic or personal dictatorship. In either case, however, industry is controlled by government.

There is, thus, a great deal at stake for our profession and for society; the public's comfort and freedom may depend on how well and unselfishly engineers, and all other professionals, serve the public as they practice their learned art. A more thorough understanding of this societal problem, and of how to cope with it is related to some precise definitions of professions and of their public responsibilities in Article 1B as well as to the development of a new professional ombudsman function, outlined in Article 4G.

These engineering society ombudsmen would be charged, not only with enforcing professional codes of ethics so as to minimize the need for governmental bureaucracies, but also to investigate, at the request of industry or government, those technological problems that affect public health or safety or material resources.

The definitions associated with professions, and a discussion of the obligations of professionals and their ombudsman function, must be treated in more detail if their relationship to the public purpose of the engineering profession is to be fully appreciated.

1B. DEFINITIONS AND TYPES OF PROFESSIONS

Professions may be subdivided into two categories; the traditional "learned" professions whose status has been accorded by public acclaim, and all others whose status has been assumed by some group of specialists. Medicine, law and the ministry achieved professional status long before they formalized their educational requirements. More recently, other professions have achieved it, like architecture; or lost it, like teaching; or almost achieved it like accounting and engineering.

Other specialists, like insurance salesmen and morticians, have formed associations which they considered professional. In general, the public has been very reluctant to confer professional status on such groups. It would, therefore, be useful to define these terms precisely.

4

Learned professions were defined most concisely by the late Roscoe Pound, dean of law at Harvard University from 1936-1956, essentially as follows:

> *A profession is an occupation composed of practitioners whose primary purpose is the pursuit of a learned art in the spirit of a public service (9).*

A more inclusive dictionary definition would describe an individual who belongs to a profession as:

> *A professional...one who engages in a calling requiring specialized knowledge, long, intensive preparation and instruction in skills and methods as well as historical, scientific, and scholarly principles underlying such skills and methods; maintaining by force of organization or concerted opinion high standards of achievement and conduct, committing its members to a kind of work which has for its prime purpose the rendering of a public purpose.*

The former Engineers Council for Professional Development (ECPD), which accredited engineering and/or technology educational programs in the United States until January 1, 1980 when it was superseded by the Accreditation Board for Engineering and Technology (ABET), defined engineering as:

> *Engineering...the profession in which knowledge of the mathematical and natural sciences gained by study, experience and practice is applied with judgement to utilize economically the materials and forces of nature for the benefit of mankind.*

The foregoing definitions, however, fail to imply all that the public expects of a professional. Such criteria were itemized by the National Society of Professional Engineers (12) as follows:

> *A professional is an individual who:*
>
> *a) is engaged in a calling,*

5

b) *enrolls in the prolonged study of*
   a learned art,

c) *has an expertise not possessed by*
   laymen,

d) *practices an art crucial to*
   society's existence,

e) *recognizes his paramount respon-*
   sibility to serve mankind, and

f) *appears to obscure his practice*
   by mystique

"The mystique arises because the professional requires a vast array of specialized knowledge to communicate with his peers and to maintain his competence by
continual study. Any one of the six criteria have been
applied separately at times...to describe a professional...Professional status, however, is accorded by society to a vocational group only if its practitioners
follow all of the criteria enumerated above for the
definition of a professional" (12). Professional status
assumed by a group of experts is seldom acknowledged by
the public.

"It would be appropriate here to review criteria
suggested decades ago for the medical profession by
Flexner (13). He proposed that the calling, or caste,
professionals regard as their profession be learned,
practical, intellectual, teachable, altruistic and unified, and yet be based upon cultural roots" (12).

All of these criteria and definitions imply the
need for practitioners to practice their art within
ethical and legal bounds. Thus, one finds that the
practice of practically all professions recognized by
the public is governed by a code of ethics. These
codes have been adopted by professional or technical
societies, and by state boards charged with the
licensing of the practitioners. Many of these codes
are lengthy documents which specify acceptable behavior
of the professional to his peers as well as of his responsibility to the public. The latter is, of course,
the most important as far as recognition by society is
concerned. The whole concept embodied in codes of ethics has been summarized most concisely by Wisely (14),
essentially as the following professional ethic:

The professional practitioner (engi-
neer, physician, lawyer, etc.) pur-
sues a learned art at all times
within his competency and in the

> *public   interest--with   integrity,*
> *honor and dignity.*

This ethic, like those expressed in most codes of ethics, implies that the practitioners, who agreed to them voluntarily, will abide by these rules of professional conduct. In most cases the organizations involved reserve the right to enforce the codes, and to discipline violators. The public expects all professions, upon which it confers such status, to discipline all of its members who are unable or unwilling to practice according to accepted standards.

Since it is the public that generally confers the status of a profession on a group, that status must be earned. Society expects those groups to use the expertise they have mastered to serve it unselfishly, to control the practice of their art by selecting and educating suitable novices, and to eliminate from their practice all who cannot master the art or who choose not to comply. A partial summary of such disciplinary actions in engineering will be discussed later in Article 7C, but is appropriate to describe the different types of professions first.

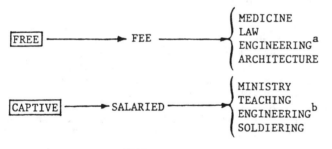

[a]Engineering consultants in private practice.
[b]Engineers employed in industry, government, etc.

FIGURE 1A - TYPES OF PROFESSIONS

Publicly accepted professions were typified as either free or captive by Marlene Dixon (3). She noted that free professionals, like law and medicine, were compensated by fees and served their clients, as Figure

1A shows. Captive professionals, like those in the ministry and teaching, are paid salaries and serve bureaucratic administrators. Perhaps no more that 10% of the engineers now practice as "free" consultants; the other 90% serve corporate or governmental bureaucracies. A century ago far fewer engineering practitioners were employees. But all professions drift from a free to a captive status with time. Today over 50% of attorneys are employees, and more physicians become salaried employees in clinics with each passing year.

Some would argue that professions must be free to be accorded professional status. They forget that the public has always considered the ministry and teaching as professional, yet they are composed almost 100% of employees and are thus captive professions. The principal criterion is whether the public believes that the peer group's primary loyalty involves service to society. Only in such cases is professional status accorded. There is a danger that professional stature, once earned, may be lost by a group if its primary loyalty shifts inwardly to its own selfish interest. Teachers are beginning to suffer that fate as they continue to unionize, to picket and to strike, sometimes in defiance of court orders.

The public service that a profession can render society is, thus, unrelated to its free or captive status. Instead, the profession's ability to serve depends upon what the group considers its primary obligation to be, whether its members are willing to sacrifice even their own economic livelihood for the public if its health and safety is endangered, and whether the professional's technical decisions are immune to revisions by non-technical, managerial authority.

1C. SOCIETAL AND PROFESSIONAL OBLIGATIONS AND GOALS

If civilizations are to exist, all elements that comprise its structure must assume or be assigned, definite obligations and responsibilities. The greatest freedom and most efficient operation will prevail when such acts are voluntary. If they are not assumed willingly, they must be assigned legally, and enforced by bureaucracies whose activities are mainly non-productive.

Representative societal elements, like corporations, governments, professions, labor unions and legislative lobbies, all play vital roles. The more teamwork they display "to promote the general welfare," the

8

less governmental restraint or enforcement is necessary, and the more productive and free that society will be. Thus, the public purpose of all elements should embrace the common goal of enhancing the general welfare as much as their talents allow.

One representative illustration of that objective is represented by the Goal's Report embraced by the American Society of Civil Engineers (ASCE) in 1973 (1). The report stated the established objective of ASCE as "the advancement of the science and profession of engineering to enhance the welfare of mankind." This report also listed the following four goals:

Goal 1   To SERVE THE PUBLIC. To provide a corps of civil engineers whose foremost dedication is that of unselfish service to the public.

Goal 2   To ADVANCE THE PROFESSION. To improve the technical capability and professional dedication of civil engineers.

Goal 3   To IMPROVE THE STATUS OF CIVIL ENGINEERS. To continue to improve the professional stature and economic status of civil engineers.

Goal 4   To IMPROVE ASCE OPERATIONS. To devote resources, organization, personnel and operation in the efficient pursuit of these goals of serving the public, the profession and the status of civil engineers.

The societal obligations of the engineering profession that further its public purpose of employing "the materials and forces of nature for the use and convenience of man" include, but are not limited to, the following:

1. Protect the public's health and safety in all engineering operations.

2. Conserve the natural environment and material resources.

3. Develop a technical expertise and maintain its competency.

4. Provide technological leadership for society.

9

5. Supply legislative advice on technological issues, describing alternate plans and listing the consequences that may follow their elimination or adoption.

6. Assume the role as the public's advocate on political issues whenever its health, safety, environment or material resources are involved.

7. Assume a role as the public's adversary against political lobbies when their activities endanger the public health and safety.

8. Discipline engineering practitioners who violate the profession's public purpose.

Most of the societal obligations listed above are self-evident. However, a few merit brief discussion.

1D. FULFILLING PROFESSIONAL RESPONSIBILITIES

Disagreements about many of the professional objectives listed above are bound to occur, even among knowledgeable professionals. It is important, however, for professionals to consider all facets of technological issues, and to advise legislatures accordingly. Generally, legislators are not technically able to prepare supportive data, since at least one-fourth are lawyers and most of the rest are businessmen. But they are charged with determining and solving public issues and must at least be properly advised. It is the responsibility of the engineering profession to structure itself so that it can supply pertinent information and at least help to determine the public's technological priorities.

But governments also have societal responsibilities. They, too, should structure themselves to act on the best technical advice available, and not to maintain the status quo for selfish political reasons. For instance, some American and Canadian railroads are poorly maintained or bankrupt today partially because national legislation did not heed excellent technical proposals on railway electrification some 50 years ago (5), and because legislatures since then were persuaded to build an ever expanding competitive network of highways, airports, and waterways. The loss in passenger and freight traffic, the taxes and subsidies that favored the other modes of transportation, and the labor legislation that condoned featherbedding forced many railroads to either defer maintenance or to go bankrupt.

10

Had all modes of transportation been integrated by national legislation and not constrained by separate bureaucracies like the Interstate Commerce Commission (ICC), the Civil Aeronautics Board (CAB), etc., their over-all operation could have been more economical, and efficient energy-wise (6). For instance, piggy-back transportation of truck-trailers by rail or container-ship might have developed sooner at a considerable saving in fuel expended per ton mile. A rational, integrated transportation system might also have checked the continual increase in truck size and weight beyond the designed load capacity of our highway system, and thus prevented its recent deterioration. The economic loss of this part of the national infrastructure will be discussed more fully in Article 2F.

Western Europe, where railroads, highways, waterways and airlines are government owned, has perhaps integrated its transportation system better. Its trucks are only half as large, its railways are well maintained, its waterways are heavily used, and its bridges last for centuries. Its passenger rail service is frequent and efficient, although its coaches and sleepers are not as spacious nor as modern as Amtrak's are in the United States. Were there no political constraints on modes of transportation in the United States, it is possible that the free market system would have integrated the transportation system more effectively than its European model.

The United States has structured itself so broad public policies are developed in the legislative branch, implemented by the executive branch, and interpreted by the judicial branch. This political system is the envy of the world. It governs an economic system far freer and more productive than any elsewhere. It is far from perfect despite the lengthy public hearings it conducts on such diverse technical subjects as transportation, pollution, energy, etc. People often confuse the purpose of these advisory discussions and the government's decision-making functions. It tries to obtain guidance from its own Congressional Office of Technology Assessment, from the National Science Foundation, and lastly, from statements made at the hearings by knowledgeable individuals or professional organizations.

The input from this last source is always voluminous, often prejudiced, and frequently contradictory. The weakness lies perhaps in the fact that the testimony on technical issues is presented by too many professional organizations. These have not been united to

11

determine the consensus of their knowledgeable members, nor to represent a block of voters large enough to command the attention of legislators and the public. For instance, at a hearing on automotive safety, television crews left as soon as Ralph Nader presented his views and were not present to hear those of the presidents of two engineering societies which were presented just after Nader spoke.

Technological societal issues are vital enough for both governments and professions to organize their operation to "serve the public" optimally. Other than to suggest changes in legislative procedure, there is little that engineers can do to alter governmental protocol. But engineers can reinforce and restructure their own professional practice to improve their public purpose.

First, they could begin by rigorously enforcing the state licensing laws, or the codes of ethics of their professional societies to eliminate from practice those few who prescribe unsafe or unhealthy designs or processes through ignorance, or very occasionally under pressure from management.

Second, the engineers could continue to restructure their formal education system by

a) lengthening its technical content for the first designated engineering degree beyond the baccalaureate to meet future needs;

b) broadening its economic and cultural base so that graduates will be better able to supervise the interdisciplinary designs of the coming post-industrial era; and

c) including instruction in management, leadership, ethics and professionalism for all graduates designated as engineers so that they can serve society more effectively.

Details of these suggestions have been described elsewhere (2,7,8) and will be discussed in Chapters 6 and 8.

Third, engineers should develop more effective and rewarding educational systems to continue their life-long learning experience so as to remain current technically. The "pursuit of their learned art" requires some postgraduate effort if they are to fulfill their public purpose. Article 6B discusses this postgraduate continuing education.

Fourth, engineering societies could aid individual practitioners, and industries and governments alike, by inaugurating the ombudsman function described previously, and treated more fully in Article 4G.

1E. CONCLUSION

If engineering is to attain full stature and to survive as a profession all of its members will have to consider their primary obligation to society to be that of protecting its health and safety, and of conserving its environment and resources. Secondary loyalties to one's employers, to one's peers, and to one's self will be important too, but should not be dominant if the public's health or safety would be jeopardized. All engineers should also be willing to discipline those of their number who design or operate unsafe systems or products. The more engineers pursue such accepted practice as individuals, the less need there will be for constraining bureaucratic regulations to collectivize society and destroy its freedom.

Perhaps, if all professionals contracted for their services to governments and corporations, and ceased serving as employees, they might be independent enough to fulfill the public purpose of their profession more effectively at all times. Whitelaw (10) argues forcefully for that ideal, but I feel that professionals can serve in captive professions as employees, provided they also fulfill their public responsibilities. Service under salaried vs. contractual status will be covered in Articles 2G and 3F.

But engineering professionals can seldom act alone! When societal forces that endanger the public health and safety marshall excessive strength, individual engineers should call for help. The professional society mechanism of ombudsman has been suggested here as one source to which individual engineers, or industries or governments, could appeal for at least a neutral confidential review.

Engineers can, if they only will, fulfill their public purpose of providing for man's material benefits while protecting the public health and safety, of selecting and educating their successors better than they were trained, and of developing the technological leadership society will need. If all who profess to be engineers will fulfill those obligations, they will indeed be professionals. Although their deeds will usually remain anonymous, these "pros" will have the satisfaction of having served their fellowmen. For that privilege they can be proud and thankful.

13

14

Public Crises and[a]
The Need for Professions

2A. THE NEED FOR SERVICE AND LEADERSHIP

The last chapter discussed the purpose and obligations of professions, whereas this one will treat the need for them. The connotations of these three terms are both subtle and distinct. Purpose relates to something that must be done and obligation infers binding responsibility, whereas need implies the lack of something required or desired. For instance, one purpose of some obscure religious sect might be to distribute free pious literature. However, there might be no need for such service if it was already available from other established sources. Who is to decide whether any sect--or profession--is needed? The public? The professionals? Who?

"One might question the need for any profession and let the rules of the market place prevail. Arguments as persuasive as 1980 sunset laws in thirty-one of the United States, the inclusion of laypersons on some state licensing boards, and the periodic reexamination for licensure have been made because some professions are viewed by an articulate segment of society as unnecessary, untrustworthy, selfserving monopolies. A century ago the playwright, George Bernard Shaw, characterized 'every profession as a conspiracy against the laity' (15). The poet, Ogden Nash, was more disparaging when he wrote (16),

Professional men, they have no cares,
Whatever happens, they get theirs.

---

[a]Presented, as part of the paper "Why Professional Schools for Engineers?" (60), at the American Public Works Association--Florida Chapter Meeting; Daytona Beach, FL; April 27, 1979.

Yet, certain *learned* professions have survived since the dawn of civilization. They must have had an accepted practical public purpose. Their practitioners were qualified as acceptable by some knowledgeable governing body. The qualification procedure was developed because the public, which relied on the professional's services, was unable to distinguish the competent from the incompetent. Furthermore, the public service rendered involved a crucial need and, at least for medicine, architecture and engineering, affected the public's health and safety as well. The governing bodies of the several professions were, and are, a hierarchy of a church, a governmental bureau, or a voluntary association of concerned professionals." (12)

Professions have always served people, either directly as does medicine or indirectly as does engineering. A small number of the professionals have exercised societal leadership as governmental officials in addition to their service as practitioners. The public needs both types of service. Society would become anarchic without government and, in our fragile technological civilization, perish without professional help. The public does try to protect itself against faulty professional or corporate service by the creation of governmental regulatory agencies, as will be discussed in Chapter 5. However, the people seem powerless to protect themselves against the unnecessary growth of these bureaucracies. Hence, governments grow ever larger and more collectivized, and personal freedoms decrease.

There is a real need for societal leadership by professionals and laymen that will restrict government activity solely to the protection of the lives, property, privacy and freedom of speech of its people. If that objective fails, and governments become ubiquitous and dictatorial, professions, as defined in Chapter 1, may cease to exist. The need for and principles of leadership will be treated in detail in Chapter 8.

Newton (178) makes a plea for such leadership when she discusses the role professionals play in positions of responsibility. She compares the sociological origins of professionals as functional or as power oriented. The former represents a bargain struck between society and the vocations, and represents a service ideal with colleague control. The power oriented concept considers professions organized to obtain social and economic advantages with ethical codes developed as a public relations necessity, and justified for the prestige the codes attract.

Newton also notes that whereas most physicians are self-employed, most engineers are employed, aspire to become managers, and are thus dependent upon profits and the business cycle. They are thus prone, as managers, to favor educational fragmentation rather than the broad technical and cultural programs needed for engineers to play a role as individuals in positions of responsibility (178). If we are "moving from the specialist who is soon obsolete to the generalist who can adapt" as Naisbitt (179) wonders in MEGATRENDS, and if we are rapidly becoming a service society where professionals become an emerging elite as Newton points out, should society then require adequate educational programs to force professions to serve the common good? Who should hold the professional accountable, and how, and for what?

Were there no professions today, the public might try to protect itself against loss due to faulty products or services by even more legal action or by extra insurance. More governmental regulatory control might be needed then to restrain unbridled, profit-seeking industrial practices. Steffens (186) pointed out that over the first century of America's economic growth such control led to two political constraints; i.e., to the enactment of the Sherman and Clayton Anti-Trust Acts in 1890, and to the formation of a host of governmental regulatory agencies beginning with the Interstate Commerce Commission (ICC) in 1887 and the Food and Drug Administration (FDA) in 1906. It was followed by scores of others culminating with the well-known bureaucracies of the 1970's listed in Table 1A. The government now regulates business and the professions because they failed to regulate themselves adequately.

The anti-trust laws endeavor to ensure competition and do prevent the formation of industrial monopolies but, unfortunately, not of labor unions. The regulatory agencies do attempt to protect "the public health, safety and welfare" by sets of regulations that are exploding at the rate of 150 pages per day! We now have more protection than we need, want or can afford. Even so, without these constraining acts and agencies, society could try to protect itself against financial loss from dangerous technology by insurance, or by contractual agreements or by the responsible practice of dedicated professionals (17). The cost of insurance would filter through the economy to the consumer. Actually, insurance does not prevent a loss; insurance merely distributes the cost over all who pay premiums. The cost of enforcing the extra contracts would spawn a plague of lawyers and clog the judicial system.

However, there are extra invisible costs inherent in the present anti-trust and regulatory enforcement. For instance, the Department of Justice (DOJ), in its zeal to prevent monopolies, sometimes regards inventions as such, rather than as a property right with a limited 17 year life. The DOJ's court actions against inventors have sometimes had a discouraging effect on industrial innovation (18, 19, 20). DOJ's actions, along with occasional shortsighted corporate policies relating to rewards for employed inventors, have been partially responsible to the loss of America's former technological supremacy to other nations, particularly West Germany and Japan (18).

The invisible costs the regulatory agencies impose on the economy are twofold. They involve direct support plus the added burden they place on industry, not only to comply with a myriad set of confusing and sometimes contradictory directives, but to process an ever expanding deluge of paper. It has been estimated (17) that if only half of this effort could be curtailed by delegating the responsibility for guarding the public health, safety and welfare to engineering practitioners, rather than to their employers, and that if the cost of that half were returned to industry, its net profit would almost triple! That profit, as described in Article 5E, would supply the capital for needed plant modernization and expansion, and perhaps restore American industry's competitive advantage in world trade.

Professions still exist today. Some are more learned and better known, and fulfill their public purpose more adequately. Among these, the ministry, the military, medicine and law have existed ever since city-states developed after crafts and agriculture had provided a surplus of food and of leisure for their elite sets of dedicated professionals to practice. Others like architecture, teaching and engineering grew out of vocations, are more recent and were less readily accepted. Yet all must have the same objective if they are to enjoy professional stature accorded by public acclaim. That objective must regard service to the public as paramount and involve an expertise unknown to laymen. Peer support among professionals, and confidence that ethical behavior is desirable, is as essential as their willingness should be to discipline those

of the group who are unable to practice within acceptable standards, or who choose not to conform. Thus,
every professional is a debtor to his profession.
Barzun (21) has intimated that democracy judges professionals by their worst examples and that public expec-
tations aim much higher than mediocrity for professional service.

The public has sustained and probably will continue supporting genuine professional leadership in this
technological civilization, just as the public's desire
for political reform outpaces legislation. If, then,
the public is as receptive to ideal professional service today as it has been throughout history, the need
for professions seems assured. However, the public is
becoming unsympathetic with unconcerned professional
behavior that culminates in unnecessary surgery by physicians, unwarranted strikes by teachers, unethical
practice by lawyers or unsafe production by industry
(22).

The public is also becoming increasingly concerned
about the risks imposed on it by technology. Cohen
notes in a forward to Inhaber's book ENERGY RISK AS-
SESSMENT (180) that "Historically, public acceptance of
technology was based almost entirely on economics...the
less costly technology was inevitably chosen...Technol-
ogies (except for nuclear energy) have been developed
by private enterprise, which had little to gain by wor-
rying about environmental impacts" until recently.
Now, risk assessment is being compared quantitatively
by measuring the deaths and the man-days lost from accident or disease by employees and by the public. A
bill was introduced in Congress by Ritter of Pennsylvania to direct the Office of Science and Technology
Policy to explore risk analysis so that it would become
part of the overall economic impact (177). The need
for leadership and teamwork by government, industry and
the profession is, thus, glaringly apparent. As technology becomes increasingly complex, the need for both
broader and more specialized professional education increases.

Professions of the future will, of necessity, also
need to offer more dedicated service if they are to
avoid being forced to revert to the crafts and trades
from which they sprung. Such professional devotion for
engineers can be achieved best if those already in
practice form a participatory unity organization capable of influencing technological legislation and dedicated to public advocacy. The details of professional
unity will be treated in Article 4D. Such an effort

would need a transformation of the engineering educational system so as to imbue the novices it graduates with sufficient professional zeal. Unfortunately, our present educational system splinters professional curricula needlessly; fails to emphasize professionalism or ethics enough, if at all; and truncates the educational effort prematurely by adhering to its four year designated programs. Those are serious charges that will be defended in Chapters 6 and 7.

2B. THE NEED FOR TECHNOLOGICAL CRISIS CONTROL[b]

Even if the practice of all professionals, whose service is crucial to the continued functioning of civilization, is virtually error-free it cannot be absolutely safe. There is an element of risk in all human undertakings even when they are routine and simple. As they become more complex and involve the services of professionals from several specialties, as is the case with interdisciplinary designs, the possibility of their failure precipitating public crises, and the need for risk analysis in the design stage, increases.

The design of the Pruitt Igoe high-rise apartment, slum clearance project in St. Louis during the 1970's illustrates this case. Family congestion and illiteracy created a social environment in it that was too dangerous for civilized occupancy. The complex was abandoned and dynamited after just two years of service. It was a *sociological*--not a technological--*failure*.

So, in a like sense, is the Aswan dam in Egypt a *biological failure*. This dam changed the ecology of the Nile River basin by impounding silt and water behind it, and by controlling the annual flooding. The lack of any fallow season downstream to kill infected snails allowed them to spread the modern plague, bilharzia.

The deterioration of the railroad system in the United States illustrates the *legislative failure* of a technological system precipitated in part by unfair taxation of its rights-of-way, and by strangling economic regulations imposed by governmental bureaucracies. Four separate agencies; i.e., the CAB, the ICC, the Federal Highway Administration (FHA) and the Federal

---

[b]Summarized from discussions on technological crises presented in Reference 23.

20

Maritime Commission (FMC) administer policies governing the movement of freight and passengers by air, rail, road and water respectively. These *uncoupled* bureaucracies, with their *unbalanced* regulations which prevent operational flexibility, constrain commerce more than would the free market system.

Crises, such as the three just cited, merely followed others which abounded when the United States was founded and the industrial revolution began. Then, as now, the crisis involved inflation, public health, public safety, resources, energy, environment, communications and transportation. The crises were developed unwittingly; although sometimes deliberately, by people in power; but more often unaccountably, by complex events.

Engineers have seldom been in power politically. Yet, their inventions have been principally responsible for changing the social fabric of civilization and for generating industrial capitalism. Significant changes in social customs seldom originated with great statesmen and generals. Such leaders ratified decisions that developed as inventions like the wheel, the zero, gunpowder, irrigation, steam engines, skyscrapers, planes, radios, automobiles, nuclear reactors and computers were accepted or marketed. Whether fewer crises would have developed if former political leaders had understood technology, or if yesterday's engineers had exhibited more leadership, may be debatable. America was born in crisis, and prevailed in crisis. Crises then, as now, were either continuous or intermittent, and politically or technologically related.

Famines and epidemics were formerly constrained geographically by slow and cumbersome transportation and travel in wagons over dirt roads or in ships over rivers and oceans. Malnutrition, especially during the winter months, was commonplace. Science and technology had not yet developed synthetic fertilizers and refrigeration to expand and to preserve local food supplies, nor immunizing serums and safe water supplies to control disease. One in five of Philadelphia's inhabitants died in the 1793 typhoid epidemic. Childhood disease immunity developed only in those who survived.

Crises in resource availability were prevalent then, as they also were in ancient Greece. For instance, two thousand years ago the demand for charcoal virtually denuded Grecian forests, and left the region a semi-desert. Coal was discovered in England some four hundred years ago, barely in time to prevent its deforestation. The production of coal, as a substitute domestic fuel, expanded substantially there after 1620, not only because it was cheaper, but because the Government of Charles I inadvertently left the mining industry unregulated for a while.

Likewise, communication systems years ago were so slow that they generated protracted crises or political isolationism. At least one major battle (1814) was fought in a war after its peace treaty had been signed.

Communication improved after the printing press was invented about 1450. Later newspapers replaced the town crier and books were reproduced cheaply enough to make universal education possible. However, the presses were also used to create crises as they churned out ever depreciating paper currencies in England (1716-1720 and 1797-1821), France (1789-1796) and the United Colonies (1775-1779). These currencies were "not worth a continental"; they inflated prices as their value declined. This process continues today, throughout the Western World at least. It provides a way to saddle future generations with monumental debts--unless they are repudiated--and enables politicians to get reelected by promising utopia. The majority of voters can enjoy and share in this spoils system until hyperinflation virtually destroys the monetary system. Engineering productivity can sometimes outrace such runaway monetary inflation when the currencies are not based upon a golden standard and are free to float--or sink.

Since engineers are frequently called upon to resolve crises, some of which, like pollution, may be their own creation, it would appear that an effort should be made to control such crises by risk analyses and by minimizing the conditions on which they are founded. Most technological crises are politically related. Hence, if engineers are to be effective, they will have to relate their practice to resource conservation and environmental protection, as well as to health and safety, and to embrace a new humanistic dimension. They will, for instance, have to improve their public credibility by assuming the role of the public's advocate and lobbyist's adversary, and by insisting that "the buck stops" on purely technical matters with engineers-in-responsible-charge in industry, as it does now with physicians in hospitals on medical matters and lawyers in courts on legal matters.

So far the engineering profession has failed to decentralize responsibility by holding the engineer-in-charge ethically accountable for his work, although so-

22

ciety has held his employer financially responsible for its product or service. Many would argue that it is virtually impossible to isolate the engineer-in-responsible charge because most engineering activities are team efforts. Although it may not be easy to pin-point failure, team members frequently do have more detailed knowledge than outsiders. Should not "the buck stop" with them too? President Truman's whimsical exaggeration did not mean that he did not delegate authority and expect accountability at every subordinate level. History's judgement of his successful administration already indicates that he did, and that he also accepted the weight of his own responsibility (176). If engineers were united and disciplined enough to accept such complete accountability for their work, the public might be convinced to transfer greater responsibility to protect its safety from regulatory agencies to the profession for the benefit of everyone.

2C. PRESENT AND FUTURE CRISES[c]

Ludwig von Mises, in his book on THE ULTIMATE FOUNDATIONS OF ECONOMIC SCIENCE writes about "The history of the West" as being "the history of the fight against the encroachment of officeholders." He pleads for a new social system based on voluntarism and suggests that the time has come "to shrink the State."

Thorstein Veblen, in his THEORY OF THE LEISURE CLASS, mentions the conflict between business and technology (on which business depends). This is a continual crisis, where business endeavors to maximize profit and designs for obsolescence, whereas technology strives to maximize production and designs for permanence.

Regulatory governmental agencies, created by Congress "to protect the public health, safety and welfare," constrain industrial production with thousands of pages of regulations and sometimes with bureaucratic tyranny. Likewise, the income tax structure, with its incentives and penalties, is both a help and a hindrance. These governmental agencies, together with business and technology, form three interacting segments of the productive process that are fueled by the flow of capital. Worldwide, legislative bodies feel they must continually expand the money supply in order to reduce unemployment. Our pump priming effort now

---

[c]Ibid, Reference 23.

requires one-sixth of our economic effort and fuels inflation. Crises continue, and expand. Elimination of
controls might depress crises.

Coupled with the interaction of agencies, legislatures, business and technology, are labor and education. These last two also contribute to current
crises; labor with its adversary stance, and education
with its permissive standards although both may be
changing now. John Chamberlain, in his book on THE
ROOTS OF CAPITALISM, believes that labor unions should
forgo their insistence on compulsory membership and
agency shop and embrace cooperative profit-sharing
principles based upon economic freedom. In Europe,
union representatives are already represented on some
corporate boards of directors. But, if labor strives
for complete security, it may become the unpropertied
regimented class described by Hilaire Belloc in his
book on THE SERVILE STATE. And unless educational systems require students to study micro- and macro-econom-
ic theory and the history of both politics *and* technology, the graduates will be destined to repeat the mistakes of the past.

Rome had its slaves, welfare schemes and monetary
inflation before it collapsed. Major civilizations
take time to deteriorate. But professionals need not
stand by idly. Those who understand the interactive
nature of the social forces involved in crises may be
better able to reverse any apparent degradation and improve civilizations if they will only lead.

Whether present crises are fact or fiction depends
upon which expert defines them. No one doubts the existence of inflation, strip mining or illiteracy. But
pollution, for instance, is debatable. E. P. Shock,
former Chairman of the National Air Quality Management
Committee, describes some of these as follows (24): 1)
Much more oxygen is being converted from water vapor in
the upper atmosphere by the sun's rays than by conversion of CO_2 through photosynthesis; 2) More CO comes
from metabolism of organisms in the soil than from
cars--the CO atmospheric content being the same in each
hemisphere despite the fact that there are nine times
more cars in the northern one; 3) The CO content is 35
ppm in Los Angeles, 220 ppm in a smoke filled room and
42000 ppm in cigarette smoke; 4) Sewage, not phos-
phates, is responsible for the oxygen depletion of Lake
Erie; 5) Thinning of bird shells is due to mercury com-
pounds, not DDT; 100 million species have disappeared
since life began 3 billion years ago, averaging 50 per
century; 6) Particulants from all of man's atmospheric

pollution for thousands of years do not equal those discharged by three volcanoes in Java (1883), Alaska (1912) and Iceland (1947).

Real crises, like resource depletion, inflation, urban sprawl, illiteracy and pollution will continue but can be minimized. Admittedly, those who are elected or appointed to lead must understand the basic elements of science and engineering on which this technological civilization is founded. But that alone will not be enough.

These new leaders in the industrial governmental and societal arenas must believe in economic freedom to develop a new social structure based on voluntarism, rather than on compulsion by government or labor. Labor must be educated to abandon its hopes of ruling society, as Samuel Gompers, one of labor's past leaders, counseled a generation ago. Good government, according to Thomas Jefferson, governs least and restricts its function to the protection of life, property, freedom and privacy.

Professional leaders in industry occupy a strategic place to initiate such progress. Their paramount interest should be the preservation and improvement of our political freedoms and of our free enterprise system. Instead, some businessmen opt for short-term profits, trading advanced western technology for questionable loans to communistic nations, thus making these socialist economies appear vigorous and viable to a gullible world. These nations, according to Anthony Sutton writing in THE WAR ON GOLD, are technical captives of the West. For instance, in 1980 Russia's merchant fleet included 6000 ships, according to the Soviet Register of Shipping, yet only 34 percent of the hulls and 20 percent of the engines were built inside the USSR. Even these engines were built with foreign technical assistance and were installed in the larger ships which they propel 20 percent faster.

Professional leaders need to educate the public so that it can make intelligent decisions competently. Socialist societies have no viable economic system that can compete with our innovative one, based as it is on freedom and incentives. Those areas of the world that have never tried our free capitalistic system remain constrained by cartels or submerged in poverty and starvation.

The free enterprise system, with all of its shortcomings, is the best that has been developed so far.

25

It can be improved by placing less dependence on collectivized technical authority and more emphasis on decentralized, voluntary professional guidance. Engineers, who are as well educated in logic as any, and better than most, could spearhead the new societal structure dedicated to public advocacy. Engineers bear a greater burden than do physicians to protect the public health, safety and welfare. Note here that paramount professional service is devoted to serving the public as its advocate. It is not an adversary relationship like that which exists between labor and industry and government.

Nevertheless, an apparent adversary relationship between technology and the public might be inferred from the jobs that machines eliminate and the crisis their loss generates. No one proposed to return to the good old days of 200 years ago when child labor prevailed, men worked a 70 hour week and women labored even more on household chores. Two-thirds of the population then engaged in food production; today only 4 percent do in the United States, and they feed half of the world! "I could point out, facetiously, that, before 1750, serfs worked 80 hours per week, died at an average age of 30, and a small elite of the aristocracy had all the welfare and leisure. Now the common man works a 40 hour work week, has a life expectancy of 70 years, the elite of the meritocracy work an 80 hour week, and the indigent on welfare have all the leisure. I believe this phenomenon is called a reordering of priorities." (7)

The temporary crises that labor saving machines create in job markets disappear as more goods become available and as more service jobs become necessary and are affordable. Ultimately, machines will produce almost all food and goods. People will have to be educated then to accept more employment in service industries and to enjoy leisure. Hopefully, schools will create an appreciation for civilization's cultural heritage instead of current TV trivia. Whether educational systems can be motivated to solve that coming crisis is questionable. Certainly, members of the learned professions can help by insisting on the pursuit of excellence in their own educational programs and practice in the emerging, capital intensive age.

2D. QUALIFICATIONS FOR ENTRY INTO PROFESSIONS

The fact that we shall have crises in the future, as we have in the past, does not mean that we should plan for everything at only a national level and there-

by keep all governmental bureaucracies intact. Nationally, we should plan appropriate societal missions,
whereas locally we should be concerned with fulfilling
such accepted missions. Progress will be surer and
swifter if we deregulate national constraints for professional service as much as possible, but hold the
professional at least ethically accountable at the local level, regardless of whether he practices as a corporate employee or as an independent consultant (104).
The important consideration in the prevention and/or
control of technological crises is to select and educate qualified professionals who regard the opportunity
to serve society as their paramount responsibility.
The discussion here will be limited to *qualifications
for entry* into professions.

In a society as free as ours, it is essential to
select professionals on merit but to avoid a cast-
ridden membership. The process is complicated because
potential novices must volunteer and yet all who aspire
to entry may not qualify.

"A 1971 report by the Carnegie Commission
on Higher Education showed that nationally
40% of all the college-age youngsters go to
college; in California 66% attend. (Of the
Commission's 19 members, seven had degrees in
law, none in medicine, and only one had an
undergraduate degree in engineering.) One may
question this need, not for post-high school
vocational training, but for traditional collegiate education. Only 25% can benefit by it
according to a study made by Hollister (26).
Successful completion of general college curricula requires an I.Q. of 110 and only 25%
of the population is so blessed. Professional curricula, moreover, require an I.Q. of
120 and only 17% of the population have such
ability. This 17% segment supplies candidates
for all of the professions as well as the top
10% of policemen, crane operators and truck
drivers, etc. and 25% of skilled mechanics.
We cannot, then, dream of sending everyone to
college unless we lower the standards. Obviously lowering professional standards would
hardly serve the public interest or protect
its health and safety.

That the professional programs of the
future will be more mentally demanding, more
technically advanced, and more multidisci-
plinary seems certain. Now the question

27

arises as to how such a change in engineering curricula of the future may be realized. Perhaps it might be possible to indoctrinate present engineering faculty so that they would not concentrate their teaching wholly on technology, but would make a conscious effort to develop professionals cognizant of societal problems. Perhaps it might. But it is well known that in academic circles it is easier to move a graveyard than to change curricula. Retreading faculty might prove even harder.

Perhaps it might be easier to develop professional schools of engineering independent of university graduate (research-oriented) schools. The objectives of such professional (practice-oriented) schools would include educating engineers to the highest technical levels but also producing leaders and decision makers for society." (27)

The details of both research-oriented and practice-oriented engineering education, and of the qualifications for entry into the engineering profession will be covered later in Chapter 6, and qualifications for the practice of engineering in Article 4E.

The IQ statistics cited by Hollister indicate that only one in six high school graduates in 1950 had the ability to complete professional curricula. Not all of the top 17% volunteer to study for any profession and cannot be forced into such educational programs in democratic societies. Those few who do volunteer have a moral responsibility akin to noblesse oblige, the inferred obligation of yesterday's societal leaders. These gifted professional novices constitute a small elite--a meritocracy--that must be groomed to practice their art with the most modern techniques, motivated to serve the public as its professional advocates, and inspired to lead society as much as their talents allow.

Ludwig von Mises, the eminent Austrian economist, while commenting on this inferred obligation said, "Mankind would never have reached the present state of civilization without heroism and self-sacrifice on the part of an elite...These people did not work for the sake of reward; they served their cause." Unfortunately, the progress they generated often occurred in spite of the interference of their peers and of the inertia of the masses. These comments merely illustrate how important it is for society to encourage and select properly qualified novices for its professions.

2E. LIFELONG COMPETENCE

Ever since professional status was accorded by the public to selected groups of specialists like physicians, professionals have spent about one-third of their life learning how to practice their art. Originally their knowledge was acquired in an apprenticeship, but later this was supplemented or replaced entirely with formal specialized education which capped basic studies.

As time passed the amount of knowledge increased, but so did man's life span. Fortunately, the third of one's life spent in qualifying as a professional increased for a while in absolute magnitude to compensate for the expansion of knowledge. Lately, however, knowledge has been accelerating at a greater rate than man's life expectancy. Man's inability to absorb this deluge of knowledge has led to ever more specialization which, as Rudoff and Lucken (28) point out, "like technology, is not inherently good...its effects depend on the degree of social responsibility that each of its developers assumes. The increased specialization has made it possible to escape social responsibilities by passing the buck." Thus, it becomes more difficult to identify who is accountable for malfunctions of definite segments of professional practice. As a result, society finds it necessary to generate, or to expand, governmental regulatory agencies whenever the malfunctions endanger its health or safety or social order or economic system.

Lukasiewicz (29), in commenting on this dilemma, noted that man's ability to *manage* natural or man-made *systems* is limited by his immediate memory span of about seven or eight items. He notes that although the number of scientific journals has been doubling every 15 years, so too has the scientific population. The volume of information per scientist has remained constant. Thus, "man faces a serious crisis"...since "his inherent intellectual ability"...is..."a fixed biological quantity." His grasp of the total volume of information is diminishing rapidly, leading to an explosion of ignorance and to an acceleration of specialization. Fig. 2A taken from Lukasiewicz's paper shows this dilemma. "The areas bounded by the circles represent the total volume of information available at time intervals equal to the doubling period D...The areas of the circular segments correspond to man's intellectual capaci-

ty"...the shaded areas to information added and eliminated, and the angle to his degree of grasp.

The absolute need for lifelong education, if any
professional is to maintain his competence after he
first becomes qualified, is obvious. Economic competition in the free market system has always necessitated
continuing education for all professionals who hoped to
maintain a demand for their services. However, economic

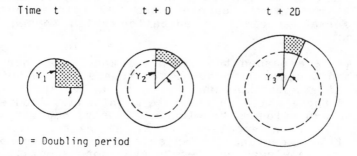

Time t   t + D   t + 2D

D = Doubling period

Circular area = Total information available at doubling periods
Shaded area = New information added eliminating the obsolete
Angles γ = Man's degree of grasp

FIGURE 2A. GRAPHIC REPRESENTATION OF THE EXPLOSION OF IGNORANCE (29)

systems that are state controlled, either totally as in
some socialistic governments, or partially as in all
other nations with their mixed economic systems, tend
to perpetuate the status quo and thus to retard innovation and the need for lifelong professional study.

Some state licensing boards in the United States
have rebelled recently at the policy of granting professional licenses for lifelong practice in virtual
perpetuity. Instead they are requiring proof of continued (up-dated) competence established by means of
periodic re-examinations or submission of credentials
which verify attendance at suitable seminars. Details
of such mandated efforts to guarantee maintenance of
competence of licensed professionals will be treated
later in Articles 4E and 6B.

Thus, the need for continued maintenance of any
professional's competence will be dictated by economic
competition, or mandated by periodic relicensure if applicable. Neither of such coercions should be needed
for a dedicated professional pursuing his learned art

30

for he, like many of his peers, derives satisfaction by advancing his art with the innovations he creates.

2F. PUBLIC'S ADVOCATE AND LOBBYIST'S ADVERSARY

Maintenance of any professional's competence is crucial if his practice affects the public's health and safety because scientific research is constantly advancing the state of his art. The engineer's responsibility is to protect the public, not only as he applies accepted or new techniques to his own designs, but whenever he observes that the plans or practices of others might endanger society or its environment, or waste its resources needlessly. It is in such circumstances that professionals should act as the public's advocate to protect its safety and resources from ill-conceived legislative efforts.

The continual efforts of the trucking lobby to increase allowable axle loads on existing highways illustrates this case. Even laymen know that when highway slabs or bridges are subjected to loads higher than those used for original designs, progressive cracking or sudden failure may result. Roads can still be repaved but bridges then need to be replaced. Both are costly. Whether gas taxes paid by trucks are sufficient to pay for such damage is debatable. John W. Snow, former Acting Assistant Secretary of Transportation, said "large trucks pay less than their share." The fact that increased loads stress pavements and bridge members above allowable design limits is accepted by all experts.

The American Society of Civil Engineers approved a report in 1974 on the SIZE AND WEIGHTS OF VEHICLES (30). It was initiated after the tragic collapse of the suspension bridge over the Ohio River at Point Pleasant, Ohio, on December 15, 1967. Forty-six lives, a score of automobiles, and the whole bridge were lost! The bridge had failed in fatigue and stress corrosion. The weight of trucks allowed to use the bridge had been increased by legislative approval from 15 tons to 20 tons over a 10 year period. This 33% increase in pulsating live load, coupled with the steady dead load, probably increased the maximum pulsating stress enough for it to exceed the endurance limit, for the bridge did fail.

The report indicated that, in 1974, "89,000 of the nation's 563,000 bridges were in critical condition" with "24,000 of the 89,000 on the Federal highway system." Of the total of 563,000, "407,000...were built

31

prior to 1935" for lighter loads than those allowed in 1974. The report also recommended that, "from an engineering point of view, no increase in vehicle size should be permitted until existing structures are put in order and the true effects of increased vehicle size and weights on bridges, pavements, and underground utilities are evaluated."

Tests have indicated that a 10% increase in axle loads above the allowable design limits will cause a 60% increase in deterioration of pavements. The American Association of Highway Officials have testified that such increased truck weights would reduce the useful life of bridges by at least 25%.

All of the facts cited above were presented by representatives of engineering societies to congressional committees five times beginning in 1969 when bills were first introduced to increase the load limits. These engineers, serving as the public's advocate and the trucker's adversary, were successful initially, for on August 20, 1975, Congress rejected the proposed 11% weight increase by a vote of 252 to 149. However, on December 16, 1975, Congress reversed itself, and shortly thereafter President Ford refused to veto the bill. Whether the fact that the trucking lobby made campaign contributions to over 100 Congressmen for the November 1975 election caused Congress to reverse itself to favor the trucking industry is a question the reader can ponder.

There have been other bridge disasters since 1967. One span of the Mianus River Bridge in Connecticut fell 70 feet killing three people on June 28, 1983. Charles Blumenthal (183), a former Maryland state legislator and a civil engineer, in commenting on the failure noted that, "Politicians, not engineers, cause bridges to fail...The bridge was properly designed for the legal maximum truck loads in 1957, but not for the greatly increased (80,000 lb.) maximums" to which state legislatures raised them "in the 1970s...The truck lobby has since argued that trucks can carry greater loads by increasing the number of axles...If the road were absolutely level and the soil pressure equal at all points this would be true...this perfect condition does not exist...As for bridges, the total load--no matter how many axles--is carried between supports of a span."

Periodic warnings about the overloading sanctioned by the legislature have been issued and have gone unheeded. The General Accounting Office Report to the Congress on July 7, 1979, on "Excessive Truck Weight"

32

recommended that Congress should amend highway legislation...to prohibit issuance of overweight permits. A 1981 National Conference in St. Louis sponsored by State Legislatures recommended that weight limits be lowered to save our roads and bridges. Blumenthal advised his "legislature of these conclusions and proposed legislation to transfer some of the financial burden to the trucking industry" but without any success.

Counter arguments by Richard Lill, Director of Engineering of the American Trucking Association (18A) stated that, "There is nothing in any of the official policies of the Federal Highway Administration (FHWA) that states that the Interstate system is to be designed to accommodate only maximum gross loads of 73,280 lb...FHWA has endorsed weight limits in excess of...(that) level." (Note: it was raised in 1974 to 80,000 lb.) Over one-half of the states had equal or higher weight limits at that time. Thus, the arguments and the data presented at public hearings are confusing. However, the public is well aware of the fact that pot holes continue to develop in pavements and that bridges continue to collapse.

The trucking lobby illustrations just cited might lead the reader to assume that there is an ever present need for all professionals, or their societies, to intercede in the public's behalf continually. Such advocacy is neither necessary nor possible. Civilization is now far too complex for such detailed vigilance. Often the legislative change being promulgated is beneficial or there would be no progress. Society must rely upon all elements which compose it to enhance the common welfare. Otherwise, too much productivity would be lost watching others. Professionals, however, must be ever alert to act on those proposed societal revisions which decrease the public's safety significantly. The professional should also remember that others who advocate changes may always be sincere even if their propositions are faulty. Public advocacy thus requires that professionals be able to communicate effectively with the public and its elected officials, and to educate both in an objective manner.

Changes in our social structure have developed ever since man first invented agriculture about 8000 B.C. Each succeeding innovation--such as irrigation, the wheel, the zero, the lever--slowly enabled the size of independent governmental units to grow from tribes to cities to nations, yet the preponderant majority of people were still farmers or lived in rural areas. The

33

advent of the industrial revolution which began less than 300 years ago, accelerated the movement to urban areas as inventions like tractors, fertilizer and power generation decreased the need for human toil and increased man's wealth and leisure.

The recent introduction of nuclear power, computers, synthetic fibers and composite materials has led to a virtual explosion of the production of material goods. The distribution of the world's goods, however, is still as skewed as it has been throughout history, for 5% of the people consume 30% of all production and 30% live in want and starvation.

The creation of this industrial cornucopia has also unfortunately generated technological side effects like pollution, urban sprawl and resource depletion, which have at times affected the public's health and safety significantly. There is grave concern among enlightened men about our polluted environment, our exploding population, and our dwindling resources. The most pessimistic critics foresee extinction of the human race a few decades hence when the polluted water and air will no longer support animal or plant life. They prophesy such doom, even though they are charitable enough to believe we will avoid atomic war.

Other critics fear the unbridled reproduction of our species will literally starve us off this planet even if we fail to multiply until there is "standing room only." Still others view our waste of material resources through planned obsolescence of the products we manufacture and the gigantic problems their disposal creates as evidence of man's thoughtless greed for profits and affluence and lack of concern for the effluent he creates.

All of these critics serve a useful purpose if only to sound the alarm. They have what they consider to be the facts to prove their thesis. But they are not always right and sometimes their timing is off. For instance, the famous Malthusian doctrine of 1798 predicted that man would reproduce faster than he could increase his means of subsistence and finally reach a state of impoverished equilibrium. But Malthus failed to foresee man's inventive ability to increase soil productivity, to improve the transportation and storage of food, or to multiply man's productivity of clothing and shelter using machines rather than slaves. But the problem Malthus recognized was real. It just took 170 years for leaders of society to take him seriously because technology bought more time for the human race to

solve its problems. Meanwhile, technology helped pollute this planet a bit more as garbage dumps and automobile junk yards vied with expanding populations for
"living room."

2G. PROFESSIONAL FREEDOM vs. GOVERNMENTAL
     REGULATION[d]

"Recent decades have witnessed an accelerating sociological and technological change. Toffler has predicted that its impact on civilization will be a future
shock (32). There can be no status quo--not for long.
If this technological society is not to disintegrate,
it will have to face a formidable array of challenges.
Leadership of all professions will be needed. Engineering appears to be well suited to participate in
that role since engineers best understand technology..." (31).

Curiously, as science and technology have advanced, political leadership in the Western world has
retrogressed. To counterbalance this deterioration,
engineers are beginning to realize that partially, at
least, their technological solutions generated problems
as well as progress, and that they must now accept an
expanded responsibility if they are to serve the public
fully. Engineers can no longer be content to solve all
problems as single entities in a complex system. They
must consider the interrelationships of other elements
of the system, presenting alternate solutions, and cit-
ing the consequences of illogical choices. For instance, in retrospect, we now see that it made little
sense to subsidize highways and airports and bankrupt
railroads.

Life is not so simple as to permit any professional to serve only one master. Hopefully, he can serve
the public, his profession, his employer, or client,
and look out for himself too. It is in this more complex situation that he generally finds himself. It is
then that the priorities of his choices get mixed up
and that his determination to look after the public interest at all costs wanes. When confronted with ethical violations of his peers, or unsafe practices of his
employer, he loses his whistle or blows it too softly.
Sometimes even this hesitation is in the public interest, for his attempt to play God may be based more on
emotion than fact. He thus avoids the quagmire encoun-

---

[d]Partial summary of Reference 31.

tered when he believes a violation exists but finds no support from his peers who reach a different but logical conclusion. It is in these gray areas that professionals, employers, and politicians need most to communicate.

One can argue as to whether whistle blowers or professionals really are needed. Do not our civil laws protect the public health and safety by holding manufacturers of consumer products and corporations offering essential services liable for their acts? The answer is yes and no. Most industries try to comply; a few of those who cut corners get caught. Collecting damages for the injured can involve costly legal battles. The public has wearied of this charade and encouraged politicians to establish the Consumer Product Safety Commission (CPSC) and the Occupational Safety and Health Administration (OSHA). The "professional" engineers in a few firms were too insensitive to potential hazards or unable to deter company management to avoid these errors. Now all firms must contend with just two more bureaucracies. The danger here is that these new bureaucracies will transform to ICC or CAB status as officials are appointed who conceive of their roles as providing protection for firms against both competition and consumer complaints (33). Would it not be less expensive--and less frustrating--for all if the technical opinions of real "pros" did prevail?

Actually, in our free enterprise system, the primary objective of industry is to maximize profits within existing legal constraints. But would not the emergence of a corps of engineering professionals within the industrial structure, whose ultimate loyalty is to the public, be just as constraining as CPSC and OSHA? Perhaps. But would they not be far less costly and generate more good will than faulty products or service?

Regardless of which constraint is best, or worst, both are with us now and both must develop greater corporate social responsibility. Manne, in discussing this topic, is concerned that "voluntary independent, corporate action...would allow industry to avoid competitive solutions" and "can only succeed if the free market is abandoned in favor of greater government controls" (34). Wallich, however, argues that, "corporate social responsibility has the advantage of shifting from the public to the private sector activities that should be performed with maximum economy rather than maximum bureaucracy" (34). This predicament suggests that corporate, government, and professional leaders

36

will have to maximize their effort to create the most good for the most people. The emerging energy crunch and the earth's dwindling resources will require better teamwork in the future.

Salvadori summed up this dilemma by saying, "The profession should be concerned with questionable morality of accepting decisions of our employer when these appear unethical or contrary to our professional integrity" (35).

No better plea for technological leadership could be voiced than that stated by the National Academy of Engineers. Its 1969 Proceedings noted that, "Leadership will be the key factor in molding the world of tomorrow. The engineering profession has the technical competence...to provide the leadership...The exercise of social responsibilities as related to that engineering objective will become paramount as man plans for the future" (36).

Forrester felt that new engineers "have moved to compete with scientists rather than crystallize a unique position for themselves as the link between the scientist and society" forcing the manager to "assume much of the role that the engineer might properly have claimed." "What kind of an engineer does the world most need?" Engineers who "can harness the...world's resources for results most beneficial to society" and who can "couple science, economics, and human organizations." Instead of such a role, "the engineer, who at one time was the educated and elite leader in matching science to society, is fast becoming just another member of the industrial labor pool" (37).

How, then, can engineers best be used in the public interest? A novel approach...originated by... Whitelaw (10) is outlined here. If engineers are to remain both free and responsible, they will have to cease being employees. They could execute 3-yr or 5-yr contracts with industries, governments, or universities for their professional services. These contracts could protect the contractor's (i.e., former employer's) privileged information better than is done now. The remuneration would not require deductions for fringe benefits since each engineer could buy his own through his professional organization. Thus, hospitalization, pensions, etc. would be truly portable. And the fixed-term contract would provide enough professional freedom so that the engineer would not feel obliged to ignore a contractor's unethical or unsafe practice. Contractors would not be obligated for severance pay and could eliminate unprofitable contractual arrangements gracefully by simply not renewing them. Thus, they would not need to worry about seniority, tenure, etc. Neither would contractors ever need to worry about collective bargaining since all professionals would act as individual consultants.

It would take time to reorganize corporate and governmental bureaucracies to accommodate the shift of its professionals from employee to consultant status. It might pay off, especially for small firms. Certainly, the public might be better served, especially if the professionals disciplined themselves as society expects them to.

Such free professionals reimbursed by fees paid by clients, contrasts with captive (employed) professionals, reimbursed by salaries paid by employers. Can captive professionals protect the public health, safety, and resources as well as free professionals? Of course, but only if they are truly dedicated to societal service. Professionals in education had that image before they unionized and went on strike. Professionals in America's Armed Forces still do have that image, devoted as they are to duty, honor, country and to remaining loyal and subservient to civilian authority.

It would take time and energy to persuade employed engineers to make a transformation from captive to free status. But once the change was made, the need for remaining technically current would be only too obvious. The benefits to themselves--as well as to society--of limiting their number to all who could qualify to rigorous practice-oriented educational standards is obvious.

3

The History of
The Engineering Profession

3A. VOCATIONAL SPECIALIZATION

The time when engineering emerged as a profession from a vocation is impossible to specify accurately. The time span selected would depend upon how the historian distinguished between the vocation from which engineering evolved and engineering itself, and upon how he differentiated between the engineer's actual practice and the obligations the engineer assumed to serve society, as defined in Chapter 1. It seems appropriate here to define the functions of those specialists who compose the modern vocational spectrum before trying to trace the history of engineering as a profession. These definitions, as assumed here, are listed in Table 3A and cover the whole technological effort.

TABLE 3A - THE TECHNOLOGICAL-VOCATIONAL SPECTRUM*

1. *UNSKILLED LABORER*

An unskilled laborer, lacking any manual or mental skills or choosing not to apply them or exercising them without much initiative, performs simple repetitive functions, assigned by a supervisor.

2. *CRAFTSMAN*

A craftsman applies repetitive manual skills to make a designed product or to render an established service.

---

*These definitions, except for No. 1, first appeared in Reference 12.

3. *TECHNICIAN*

A technician applies <u>proven industrial techniques</u> under the direction of an engineer or scientist. (Example: draftsman, machinist, etc., who might be certified.)

4. *TECHNOLOGIST*

A technologist applies <u>established scientific methods</u> under the direction of an engineer or scientist. (Example: chemical analyst, computer programmer, etc., who might be certified.)

5. *ENGINEER (PROFESSIONAL) (E.P.)***

An engineering professional applies accepted, and develops new, engineering materials and principles to design, develop, produce or operate products, structures, machines, or systems, and/or to manage these for man's use and convenience, being ever mindful of the public's health, safety, resources and environment and of the appropriate standards of ethical conduct.

6. *PROFESSIONAL ENGINEER (P.E.)***

An engineering professional is licensed by the state to practice whenever public health and safety are involved and the practice is not exempted. He is authorized and may be required to apply his state seal to his designs.

7. *ENGINEER-IN-CHARGE***

An engineering professional in <u>responsible</u> charge of a project who might also be

---

**Engineers may be certified, registered and/or licensed. P.E.'s must be licensed. Scientists are seldom licensed or registered.

TABLE 3A (continued)

legally <u>accountable</u> for his work.
should be "where the buck stops."

<u>8.</u> *SCIENTIST*

A scientist discovers new scientific principles and materials, and monitors established techniques and processes.

Unfortunately, all of the terms itemized here have been used ever so loosely by the public, government and industry that they must be defined accurately if individuals are to be qualified for any of them.

The fact that the term "professional engineer" (P.E.) was defined separately does not imply that engineers in the other two categories are not professionals. Many, although by no means all, are professionals even though they may not be "registered." The criterion distinguishing the engineering professional (E.P.) from others of his peer group is his willingness to regard the protection of the public's health and safety as paramount, and of his willingness to uphold the code of ethics of the engineering society to which he belongs.

The title professional engineer has been reserved by state licensing boards for those engineers it licenses to practice. However, the practice is exempted if it does not involve public health and safety, or if the engineer is supervised by a "registered" engineer or if the engineer is an employee of any governmental agency, public utility or industry engaged in interstate commerce. These last three exemptions are usually referred to as the "industrial exemptions" (12).

All of the vocational specialists just enumerated could also be grouped according to crafts like carpentering and welding, or by occupations such as those for electronic technicians and computer programmers. The engineer or scientist himself might also be classified by his own specialty, if he is a civil engineer, nuclear physicist, etc. These specialties have been proliferating and developing since the dawn of history, but they reached enough of a critical mass to trigger the industrial revolution about 1750.

It is not the purpose of this treatise to narrate the history of engineering, for innumerable excellent sources exist already. Rather, the aim here will be to first trace the evolvement of engineering from isolated innovations to the development of related crafts and vocations, and finally to the craftsman's substitution of scientific principles for trial and error techniques. That is when we might consider engineering to have been born even though it may not have been recognized as a profession until several millennia had past. But do just the few principles of physics, chemistry and mathematics known to the builders of civilization's first irrigation projects in Mesopotamia about 5000 B.C., or to the builders of Egypt's pyramids about 3000 B.C., suffice to identify these artisans as engineers or architects?

They were knowledgeable about elementary metallurgy, hydraulics, mathematics and construction practices. The Egyptian engineers were so revered for their ability that they were advisors to royalty and given the title chief of works. One of them, Imhotep, was elevated to the status of an Egyptian god after his death. Beakley (38), in commenting on their ability, warned of their weakness as follows:

"Although the skill and ingenuity of the Egyptian engineers were outstanding, the culture lasted only a relatively short time. Reasons which may account for the failure to maintain leadership are many, but most important was the lack of pressure to continue development. Once the engineers formed the ruling class, little influence could be brought to bear to cause them to continue their creative efforts. Since living conditions were favorable, after an agricultural system was established, little additional engineering was required. The lack of urgency to do better finally stifled most of the creativity of the engineers and the civilization fell into decay."

It is quite likely that the availability of slave labor was equally discouraging to technological innovation. On such substance do civilizations crumble or revolu-

tions evolve. Beakley's message related to the
societal obligations the specialists adopted voluntarily. If the services of these specialists benefited
only royalty or the gods, and were not related to improving the material comforts and safety of their
fellow men, their practice would not have been
considered professional as defined in Chapter 1.

The fact that these early engineers formed an
elite group would not have prevented them from being
regarded as professionals if they had only chosen their
primary loyalty as being to the public and not to their
employer (royalty). The latter loyalty was important
in its time for construction of Egypt's irrigation
works and pyramids did provide employment for the masses, whether they were slave or free.

Although many historians do regard these early
builders as engineers, the *title of engineer* is considered by both Beakley (38) and Mantell (39) to have been
first indicated about 200 A.D. when historical manuscripts referred to an invention as an *ingenium*. A
millennium later an operator of a warlike device was
called an *ingeniator*. Thus the engineering title relates more to a term like ingenuity than to an engine.
The European spelling of *ingenieur* is more appropriate
than the English spelling of *engineer*, for the former
relates solely to *creativity*, whereas the latter mainly
to *operation*. The English speaking public fails to
differentiate these functions and conceives of engineers as those who design intercontinental missiles, or
operate power systems, or drive locomotives or adjust
hotel air conditioners.

If the English speaking *engineer* is ever to escape
from such a mixed vocational status and have the public
conceive of him as a creator rather than as an operator
of technical systems, he could eliminate confusion by
adopting the European title of *ingenieur*, or the
Spanish one of *ingeniero*. In fact, Spanish speaking
engineering graduates shed the title Senor (Mister) and
substitute Ingeniero at their commencement. It would
be a simple matter for English speaking engineers to
accomplish this change. All that would be required
would be for the engineering societies to change, by
action of their boards of direction, the spelling of
only this word in their constitutions, publications,
etc., and for engineers to address themselves accordingly.

The title might also be changed for worldwide usages to *ingenior** "to emphasize the practitioner's *ingenuity* for designing products and managing systems, and to distinguish his function from those of technicians who help make and operate them" as Samson and I suggested in a futuristic, historical fantasy. We also imagined that engineering unification by 1992 would lead ingeniors worldwide "to address themselves as *Genior,*** thus...paralleling the title of doctor used by physicians internationally." (181)

These suggestions are proposed here merely to stress the paramount public function of the engineer to be that of protecting the public's safety and resources as he creates new products or systems to serve society. The change, particularly if he were addressed as Genior instead of Mister, would always remind him and the public of his primary societal obligations. The suggestion is not made to establish an elitist group per se, but to form a nucleus of specialists dedicated to serve mankind unselfishly.

3B. ARTISANS, APPRENTICES AND GUILDS

The evolution of engineering as a technicological function and as a profession was alluded to in Article 3A. Perhaps the public acceptance of engineering as a profession should be considered as occurring after the Industrial Revolution began about 1750. It was about that time that engineering schools and technical societies were first founded. Their origin will be covered in detail later in Chapters 4 and 6.

It seems appropriate here to dwell briefly on the early organizational efforts of engineering as a profession by discussing its relationship to artisans, apprentices and guilds which comprised early facets of the technological spectrum.

The history of technology is essentially the history of man as a creator of new tools, products and systems (40). He is the only animal capable of innovating enough to modify his environment and has lately

---

*Ingenior (pronounced In-je-nior' or In-jen-yor') relates to ingenious and ingenuity.
**Pronounced Gee-nior or Gen-yor. Thus it resembles the Spanish word Senor.

44

even learned how to annihilate it if he so chooses. Other species only produce artifacts like hives, nests, and dams, or use sticks as levers. Man's success as an innovator was based both upon his desire to improve his living standard, and upon the incentives which existed in the socioculture in which he lived. His material progress began to accelerate soon after the invention of agriculture and the development of irrigation provided a surplus of food so that all could eat, of time so that specialists could develop, and of leisure so that an elite could contemplate. These specialists were soon composed of artisans who provided the critically needed goods and services that the ordinary man could not produce, and of public servants needed for governing the city-states to provide for the common defense, to maintain order, and to distinguish custom from law. The several specialties were soon characterized as vocations, a few of which like the military, the ministry, law and medicine began to assume the status of professions. All of these vocations were, thus, a function of the existing social structure. Many vocations appeared, flourished and vanished depending on their need. The scientific principles upon which they may have been based may not have been understood. Their technology was sometimes regarded as a mystery, or required extended periods of time to learn. In either case apprentices were initiated into the art like priests, or groomed for practice like physicians.

The duration of apprenticeships decreased as periods of formal education in the specialties were prescribed after appropriate schools and programs were developed. A ministerial elite was educated in Egyptian schools as early as 3000 B.C., and Greek physicians at the Temple of Cos in the 4th Century B.C. Apprenticeships survived as the sole method in engineering education until 1747 when the French Ecole National des Ponts et Chaussees opened. The history of this educational effort will be expounded in detail in Chapter 6.

Apprenticeships in most vocations, which are not based upon the pursuit of a learned art, is still the only way to learn some specialties like brick laying. There always were qualifying tests which marked the successful completion of the apprenticeship, although the constant evaluation of the apprentice's experience was usually monitored. Today some vocational graduates are examined for certification by state registration boards, as is the case for plumbers, auto mechanics, etc.

As time passed artisans who composed particular vocations found it advantageous to unite into guilds or craft unions. Labor unions, as such did not materialize until the 19th century, but guilds existed as early as the 5th century A.D. for Grecian shipowners, and for trades in *collegia* or *corpora* of Rome and Constantinople. These were of two types; i.e., social or religious associations organized for mutual protection from thieves, for the distribution of alms, for provision of burial services, and for celebration of feasts and in processions.

Two other types of guilds formed the early origin of the engineering profession. These emerged in the 9th and 10th centuries A.D. The first was the *merchant s guilds* or association of traders engaged in international commerce. The traders united to protect their caravans, and established monopolies in towns and city-states by gaining appointments to their governing bodies. The trader's interest in such commerce was soon extended to the manufacture of the artifacts and cloth they sold. However, the artisans who produced these goods became even more specialized and began forming their own *craft guilds* about the 10th century.

Merchant guilds thus *regulated the* local *economy* and *craft guilds controlled the* local *industries*. Both were interested in creating monopolies of trade and production, but the craft guilds were also dedicated to guaranteeing good materials, workmanship and working conditions. Production standards were established and innovations forbidden in the interests of uniformity. The workmen consisted of the master craftsman who owned the productive establishments, journeymen who were employees, and apprentices who served up to seven years to learn their trade under an indenture or contract, and who were obligated not to marry, run away nor frequent taverns during their apprenticeship. Some apprentices paid fees but were provided with room, board, pocket money and a general education in addition to learning their craft.

Enterprising masters sometimes avoided training apprentices by hiring only journeymen and assigning them to specialized, repetitive tasks to cut costs. Journeymen were frequently excluded from full guild membership, and were known to strike and riot for higher pay and better working conditions but never operated solely in labor unions. The wealthy employer class in

46

Flemish and Italian cities tried to prohibit separate journeymen's associations by law.

As commerce grew and production expanded, the capital needed for journeymen to start their own business increased until the amount needed virtually exceeded their resources. Manufacturing shifted from urban craft guilds to rural cottage industries. Rich merchants became the new capitalists and entrepreneurs, master craftsmen became foremen and journeymen the workers. Thus, although some guilds were still being formed in the 17th century, the guilds reached their zenith in the 16th century and had probably outlived their usefulness a century or two earlier than that. By 1791 in France, 1814 in England, and 1864 in Italy, craft associations had been abolished legally.

The reader should note that although some of the societal obligations adopted voluntarily by craft guilds were definitely professional, others like trade monopolies and innovative prohibitions did not enhance the public welfare. Free market capitalism based on giant transnational corporations, and engineering societies based upon participatory membership of individual professionals, finally emerged and developed to their present state. Both emerged shortly after the Industrial Revolution began about 1750. The history of the engineering technical and professional societies will be covered in Chapter 4.

3C. MANUFACTURERS

Artisans have made artifacts since the dawn of civilization. Their productivity increased as the tools they used improved, but it was not until the introduction of rotary motion using the potter's wheel, the bow drill, and the pole lathe itself that productivity began to accelerate. Before then, manpower had been supplemented by that of domesticated animals and sails for ships, and the screw, pulley and lever, although manpower was still used on tread mills which powered cranes, and drove scoop mills to raise water for irrigation. Many water mills were operating by the end of the Roman era.

Manufacturing during the dark ages (AD 500-1500), as distinct from construction of buildings and roads during that era, was confined to pottery, flour, cloth, furniture, wine, jewelry, cosmetics, etc. These items were produced by artisans, and then by guilds when resources like capital, materials and labor were required in greater quantities. Later, urban markets developed which encouraged the innovative application of inventions enough to support a mercantile economy. What had been an agrarian society based upon a subsistence economy was changed into an industrial society based upon trade and town life.

This change was influenced mainly by the transfer of technology which was related to the travels of traders, to the sale of their artifacts, and finally to the printing of technical manuscripts. The Chinese, for instance, had mastered the art of metal casting fifteen centuries before Europeans did, and were the first to burn coal, make mechanical clocks, and invent gun powder, the compass, silk and printing with movable letters (42). But the Chinese did not apply this inventiveness to their industry. Like most East Asians, they endeavored to master themselves to adjust to their natural environment. Westerners, conversely, endeavored to master their environment to optimize their material well-being.

Medieval technology was transferred throughout Europe and the Near East as well as with China. But Western technology emerged so rapidly from 1500-1750 A.D. because of the different philosophical objective of its civilization, and because that created a social environment conducive to invention as well as to innovative applications.

The industrial innovations marketed from AD 500-1500, like the spinning wheel, gunpowder, cast iron, printing, clocks and windmills, were as important to the development of the cottage textile industry and small scale custom manufacturing, as were later innovations like mechanized textile mills (1719), steel (1744), and the steam engine (Newcomen's in 1712 and Watt's in 1769) to the creation of mechanized urban manufacturing industries. This transformation was, however, related also to social changes like the religious Reformation, the scientific Renaissance, the accumulation of capital, and the emergence of the nation-state. The mercantile economy of the middle ages was quickly replaced by an industrial revolution beginning in England (1750) and coinciding with two political ones in America (1776) and France (1789). All three strove for the dignity of man and a unity with nature, endeavoring to free him from want and oppression. Progress was greatest in the United States because the founders of its social institutions--men like Franklin and Jefferson--used a wide time horizon directed toward the future. But, before this outburst of technology was unleashed, both the merchant and the craft guilds had been successful in persuading governments to create monopolies which retarded the progress of technology as much as did slavery. Rougier (42) in his book THE GENIUS OF THE WEST, notes this constraint as follows:

"The medieval corporations...blocked all innovation. In 1643, the British Privy Council, not content with simply refusing a patent for a revolutionary knitting machine, ordered its destruction. In France, enforcement of the prohibition of the import of printed calicos cost the lives of thousands of people. In the town of Valenciennes alone, 77 people were hanged, 55 broken on the rack, and 631 sent to the galleys for the crime of trading in these articles. 'The regulations are so rigorous... that officials and clerks at the ports of entry may legally strip any lady venturing to wear in public a gown made of linen.'

To enforce these complicated regulations most European countries maintained an army of inspectors...The corporations, and the monopolies created by the state, had the same crippling effect on invention and innovation as did antique slavery...

The Industrial Revolution required the freeing of commerce and industry from the regulations of the corporations and monopolies. This was done in France...'As of April 1st' (1791), the law declared, 'a citizen will be free to exercise any profession or practice any occupation.' In England it was not until 1813 that the Statute of Artificers of 1563, which regulated the numbers of workers and apprentices, was repealed. The field was free for the operation of market forces which... would in the course of a century and a half transform the world" (42).

The preceding discussion on constraints imposed upon manufacturing by guilds and their monopolies contrasted with the expansion of production created by inventions and their innovative applications from 500 to

1500 A.D. Many of the inventors could definitely have been classified as engineers since they endeavored "to use economically the materials and forces of nature for the use and convenience of man." But they had no organized profession to which they could belong during that millennium, or for several more centuries, and which was dedicated "to protect the public health, safety and welfare." But economic monopolies were not the only medieval constraints restricting productivity; human slavery and oppressed labor of free citizens did so, too.

3D. HUMAN VS. MECHANICAL SLAVES

The forgoing critique of artisans, guilds and small scale manufacturers failed to mention much about whether the labor was slave or free. In early civilizations like the Egyptian, Grecian and Roman, much of the common labor used to build the irrigation systems, temples, roads and aqueducts, and to produce the agricultural products was undoubtedly heavily dependent upon human slaves, or upon their economic equivalent as serfs. Hazelitt, in his book THE CONQUEST OF POVERTY, noted that, in the early Roman Empire (146 B.C.) there was one human slave for every five free citizens. By 235 A.D. this proportion had changed to three to one (43). This fifteen-fold increase in the ratio of slave/citizen was due to the successful conquests of Roman legions, and to the Roman administrator's ability to assimilate foreigners as equals. Actually, the practice of enslaving war prisoners was a humane improvement of the earlier custom of killing all male prisoners.

As the proportion of human slaves grew, there was insufficient employment available in the rural areas for free Romans, most of whom tilled the soil. They drifted to the cities where, as their number grew, the government found it necessary to place many on welfare, to inflate its monetary system, and to build coliseums where the people were entertained enough to vent their frustrations so as not to revolt. Romans never did revolt; they were ultimately conquered by barbarians from the North by 476 A.D.

One could draw an analogy between the increase in energy per free citizen produced by Roman expansion to that developed by the industrial revolution 2000 years later. Where every free Roman was supported by the equivalent of three human slaves in 235 A.D., today

every American is served by the equivalent of 200 mechanical slaves. In each case, the shift in toil from
freeman to slave produced a surplus of farm labor which
drifted to the cities. In 1776 three-fourths of America's population were farmers. By 1976 only 4 percent
were, thanks to the mechanization of the agricultural
industry.

The surplus American farm labor drifted to the
cities as these became more industrialized. However, a
sizeable percentage was always unemployed. These had
to be supported by welfare, monetized by deficit spending. Like their Roman counterparts, they too became
frustrated. Stadiums were built to entertain them.
Hopefully, they will not revolt. But will their frus-
tration generated by their lack of employment or un-
wholesome social environment destroy their work ethic,
and soften them up enough so they will trade their
freedom for a stable social order? Historian Arnold
Toynbee observed: "Freedom is expendable; stability is
indispensable" for a civilization to survive (44).

No single profession can assume credit for, nor be
blamed for, the current state of Western Civilization,
yet all are involved. The absolute gains in technology
far outstrip the relative losses in education. Rougier
commented on the former as follows (42):

> "The skilled worker of today--in terms of
> his food, clothing, entertainment and cultur-
> al formation--lives a life a master craftsman
> of the days of Louis XIV could not have ima-
> gined. The peasant of Western Europe and the
> farmer of North America bear no resemblance
> whatever to La Bruyere's savage beast, or to
> Rousseau's tiller of the soil. In five gen-
> erations the working man has gained more ma-
> terial comforts than during the previous
> twenty-five centuries."

Yet, despite these tremendous material gains, and the
fact that elementary education is now virtually universal in much of Europe and North America, far too few
people have been taught to appreciate their cultural
heritage or to enjoy their leisure. If those who enjoy
the bounty of Western Civilization are to keep their
economic and political freedoms, they will have to be
educated to appreciate, to understand and to defend
them. But who is to educate whom? This is the great
challenge of the future for all professionals, and es-

pecially for those scientists and engineers who understand this technological civilization. The ways in
which they may be able to supply this aspect of societal leadership will be treated in Chapter 8.

The gradual elimination of human slavery was due
mainly to the moral and social revolution spawned by
Christianity. Its objective was not to reform society
but to prepare people for penitence and heaven (42). A
deceased, earthbound slave was to become a freeman of
the Lord, whereas a deceased free man was to become a
slave of Christ. All men were thus of equal dignity.
This belief rehabilitated manual labor. Greek scholars
had considered it ignoble, and the practice of engineering still suffers from that stigma. Although
Christian teachings were successful after two millennia
in freeing all human slaves from bondage and transform-
ing them into serfs or tenant farmers, it was not until
the industrial revolution gained momentum before common
labor was freed from back breaking toil.

One might argue that the craftsmen who served as
the engineers of their day did improve the welfare of
all workers, and so were acting professionally. However, public recognition of engineering as a profession
did not begin to emerge until the 20th century, and has
not yet attained the stature enjoyed by medicine, law
and the ministry. It is also likely that these early
engineers did not consider the protection of the public's health, safety and welfare as their paramount responsibility.

3E. THE INDUSTRIAL REVOLUTION

During the Middle Ages technology advanced slowly
but substantially as artisans and guilds improved their
crafts. Its pace quickened about 1500 with the emergence of nation-states, the scientific revolution, the
Protestant Reformation and colonial exploration. Prior
to 1750 many inventions never reached the innovative
stage because suitable materials, production techniques, skilled labor, capital or social acceptance
were lacking. For instance, Leonardo da Vinci (1452-
1519) conceived of airplanes and submarines but they
were too advanced when he sketched them, and they met
no urgent social need then.

It was the shortage of labor and unprecedented expansion of power supplied by engines that really triggered the Industrial Revolution, and that continues to

accelerate it even today. Table 3B itemizes a few of these developments. None of them could have materialized without the substitution of coal for charcoal, the source of which was fast being depleted in 1700, and of mineral oil for whale oil. Nor could progress have been very substantial without the use of iron to build power driven machine tools.

TABLE 3B - STEAM AND GAS ENGINES

1712	Newcomen steam engine (atmospheric)
1769	Watt steam engine (condensing)
1804	Railroad Locomotive
1818	Stirling steam engine
1859	Lenoir gas engine
1878	Otto cycle gas engine
1884	Steam turbine--electrical generation
1885	Automobile (Daimler and Benz)
1892	Diesel engine
1930	Jet engine
1955	Nuclear powered electrical generation

It is also important to note that industrial progress flourished mainly in Western Europe, and particularly in North America, where political freedom and free trade were guaranteed by law. It flourished also because man was ingeneous enough to produce metals like alloy steels, aluminum and titanium; and to use fuels like coal, oil and uranium. He first developed the most accessible sources, such as mineral outcroppings, but soon was forced to dig deeper shafts for ore and drill deeper wells for oil. The costs in money, or energy, increased and will stop when more energy is expended to mine and to process fuels than is available in their finished form. Thus the world may never run out of such materials; they will just be too expensive energy-wise to mine unless, of course, recovery techniques become more and more efficient.

No one can foretell how long any natural resource will last because mineral extraction is constantly being improved, and because more deposits continue to be discovered. Nevertheless, it must be apparent that our present civilization is living on the inheritance bequeathed to it by nature which it stored and saved over millions of years. As society continues to use fuels like oil and coal, it degrades such energy into less available forms like heat. Some work is extracted but much heat escapes into the universe. It cannot be recovered. The measure of this energy unavailable for more useful work is called entropy. Similarly, minerals which are extracted from their ores also require energy. Although they may be fabricated into useful products, more are later scrapped than are recycled. Rifkin (45), in his book ENTROPY, discusses the world chaos that must result as this waste continues. That constraint will be discussed later in Chapter 5.

Ultimately, engineers will have to learn to harness solar power for energy and explore the heavens for minerals. Perhaps their greatest challenge will be to educate governments and people to support such long term projects, rather than the short term solution politicians prefer because it favors their chances for re-election. The need for such professional leadership will be treated later in Chapter 8.

Engineers, when they reflect on the gains made in the Industrial Revolution so far, should be optimistic about the future. Whole new technologies are being invented now as civilization moves into a new age founded upon instant communication with one's professional peers, efficient information retrieval, composite materials, microprocessor applications, genetic engineering, etc. Yet, engineers should remember that invention of a new device or system is useless unless it is applied, innovated, and marketed.

The innovative process today depends upon social need and acceptance of the invention, on the availability of suitable materials and capital, on marketing techniques and on legal constraints. Any industrial progress is intimately related to economic factors, especially in a free enterprise system. As these fluctuate, they tend to cause the innovations to cluster time-wise around favorable economic environments, even though the inventions might have been discovered at more regular intervals.

Some economists believe that the Industrial Revolution triggered a quantum jump in productive capacity that began about 1750 A.D. with the invention of steam power and that has continued uninterruptedly ever since. Others believe its continuation has been manifested by a series of perturbations based upon the introduction of each new group of technologies. Evidence to support the latter theory is gaining favor lately. These economic perturbations represent the cyclic variations in production of manufactured goods, steel, coal, food, etc.; and in the variation in wages, interest on invested capital, savings, prices, etc. Kondratieff (46) discovered these cycles as having a fundamental period of about 50 years that began about 1790. He compared various economic indicators such as prices, interest rates, bank deposits, wages, coal and pig iron production, cotton acreage, etc. All began to rise about 1790, 1850 and 1895, and began to decline about 1814, 1875 and 1915. These data have also been integrated into the U.S. wholesale price index and plotted in Figure 3A (47).

This long wave economic phenomenon was first published in 1925 (46). That paper merely presented facts to show the capitalist economies, from the Industrial Revolution onward, followed a regular long cycle, going from boom to bust to boom about every half century. Later in 1939, Schumpeter (48) published a monumental work on business cycles which expanded the Kondratieff Wave Theory, interpreting each new upswing as being caused by a new group of technologies. The first was based on widespread use from 1780-1842 of Darby's discovery of smelting iron ore and coke (1709) and of mechanization of cotton mills (1719). The second was caused by generalized use from 1842-1897 of steam power in plants and railways and by use of wrought iron (1784) and Bessemer steel (1856). The third wave for electricity, chemicals and autos began in 1898 (48).

Schumpeter distinguished between invention and its innovation--or application. Thus, each innovative wave caused markets to expand until they were saturated. Then profits fell, unemployment rose, and the wars which followed might have been related to a struggle for markets. But necessity, being the mother of invention, provided another burst of innovation. The demise of capitalism was thus postponed again and again, and did not collapse as Karl Marx theorized it would. He reasoned that capitalists would amass private property and deny workers their share until they revolted. Stalin did not welcome the refutation of Marx's theory and, in 1930, banished Kondratieff to Siberia where he died.

Mensch (49) cites the way hundreds of innovations peak periodically, as they have in 1764, 1825, 1886 and 1935. The next should peak in 1984. Hall (50) feels that Mensch's "wagon-train effect" was hinted at by Kondratieff and specified by Schumpeter. Inventions do occur spontaneously and frequently simultaneously, but innovations bunch up as Ogburn notes in SOCIAL CHANGE

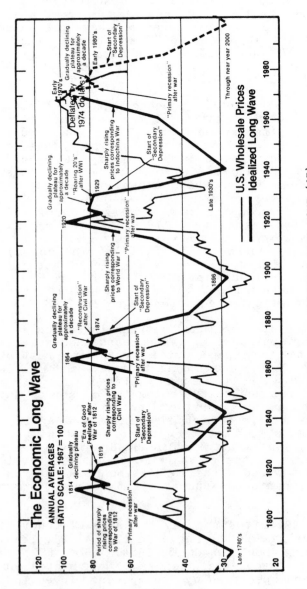

FIGURE 3A - THE ECONOMIC LONG WAVE (47)

Reprinted with the permission of the author, Julian M. Snyder, and
the publisher, International Moneyline, New York, NY

56

(51). He lists 148 inventions and scientific discoveries made independently by two or more persons which were taken from ARE INVENTIONS INEVITABLE? A NOTE ON SOCIAL EVOLUTION, (182). A few examples are: Calculus by Newton (1671) and Leibnitz (1676); Organic Chemistry by Boerhave (1732) and Hales (1732); Photography by Daguerre-Niepce (1839) and Talbot (1839); Telephone by Bell (1876) and Gray (1876); Centrifugal Pumps by Appold (1850), Gynne (1850), and Bessemer (1850). Many previous discoveries occurred simultaneously because communication was then far slower. Discoveries may continue to occur simultaneously in the future because of the increased volume of research and imperfect information retrieval (see Figure 2A). The locus of innovation has shifted geographically from England, to Germany and the United States, to the U.S., and now to the U.S. and Japan. The reader should note that as each nation became an innovative leader, it became an economic leader, also.

Schumpeter also suggested that two shorter waves could be superimposed upon the long Kondratieff. These were the Juglar and the Kitchin waves. The comparison of their relative amplitude and period to the Kondratieff and of their perturbative nature are noted in Table 3C (48), and the equation idealized in Figure 3A (47). The equation could be fitted to the wholesale price index plotted in Figure 3A using the theory of least squares. It should be noted, however, that not all economists agree with Schumpeter or Kondratieff (52). Many factors besides prices affect cyclic booms and busts. Whether equations fitted to existing indexes can predict the future reliably is questionable.

TABLE 3C - ECONOMIC WAVES (48.)

Wave	Relative Amplitude	Period = t	
		Months	Years
Kondratieff	18	684	57
Juglar	3	114	9.5
Kitchin	1	38	3.17
Resultant*			

$$*y = 18 \cdot \sin \frac{2\pi t}{684} + 3 \cdot \sin \frac{2\pi t}{114} + 1 \cdot \sin \frac{2\pi t}{38}$$

From BUSINESS CYCLES by Joseph Schumpeter (c) 1939
Used with permission of McGraw-Hill Book Company

3F. PROFESSIONAL PRACTICE AS CONSULTANTS OR AS EMPLOYEES

Engineers were first employed (3000 B.C.) in military operations constructing fortifications, roads and bridges, or by governments building temples, amphitheaters and memorials like the pyramids. They needed a sound knowledge of current scientific theory and the art of construction, as well as an understanding of the uses of natural and of societal wealth. Thus, early engineers served as employees in the existing monarchies.

Simultaneously, artisans plied their crafts either as entrepreneurs, or as royal employees--or even as slaves. Whether their work should be considered as engineering depends upon assumed definitions. As time evolved, and particularly as the Industrial Revolution unfolded, their practice became ever more technical. It definitely assumed engineering stature as machine power developed, and as factories replaced the cottage industry of the late Middle Ages. The practice could assuredly be considered as professional when an innovation like interchangeability of parts was developed by Eli Whitney for the mass production of U.S. Army rifles in 1798, or another innovation like assembly line production was introduced by Henry Ford to produce the Model T car about 1905.

It took a millennium for military engineering to shed some of its public functions so as to create the profession of civil engineering, and to found the first engineering schools. Many of the 19th century civil engineers in the United States were graduates of its Military Academy at West Point, or of its earlier French version which will be described later in Article 6D. Until 1800, most of the work was confined to surveying, and the construction of roads, wooden bridges, and water works. Shortly thereafter, the engineers were more concerned with the construction of machines, railroads and iron bridges. The advent of the steam engine and power driven machinery splintered civil engineering, forming the specialty of mechanical engineering. This splintering was followed by the formation of electrical and chemical engineering as electricity and chemical processing were introduced. Many of the engineers practicing in these non-military specialties served their clients as consultants or founded their own firms. Names like Squire Whipple, John Roebling, George Westinghouse, Eli Whitney, Henry

58

Bessemer, Thomas Edison, Alexander Graham Bell and Charles Goodyear come to mind.

Gradually, as industries and governments grew in size, they found it advantageous to establish their own (captive) engineering staff. The proportion of employed engineers grew from a small percentage in 1800 to over 90 percent today. As employees, they were not as free professionally as were those practicing as consultants. Both, no doubt, endeavored to hold paramount their professional obligation to protect the public health and safety, but at times the employed engineer's technical decisions were overriden by nontechnical management (53,54,55). These cases happened infrequently, yet often enough since the first code appeared in 1912 to cause codes of ethics to include sections requiring engineering society members to refer such cases "to the proper authority."

The proper authority was an ill-defined, amorphous mass of public, corporate and professional society personnel. Frequently, the employed engineers concerned with specific cases failed to take action for fear of losing their jobs, or because they felt "higher management" had assumed responsibility. It was not until the 1970's that engineering societies began to define and to provide service as ombudsmen on a confidential basis to assist those of their members whose employers were involved in engineering activities which might endanger the public health and safety. Details of ombudsmen's roles will be covered later in Article 4G.

Although engineering societies should expect their members to uphold codes of ethics, the societies should be expected to help their members when they are doing so, particularly if the public health and safety is endangered. It would be impossible to investigate trivial complaints, but certainly those legitimate ones emanating from the engineer-in-responsible-charge should be considered, regardless of whether he serves a corporation as a consultant or employee. Should not all employed engineers be held ethically accountable for their practice, just as their employers are held financially responsible for their products or services? Consultants, of course, must expect to shoulder both burdens.

The point to emphasize here is that "the buck should stop" at least with the project engineer-in-responsible-charge. Were this the case, the engineer-in-charge would not then allow himself to be pressured into transferring his accountability or responsibility to non-technical management. Less confusion might then exist as to whether he, his employer or a government agency was at fault if some product or practice with which he was associated might endanger the public.

Engineers, whether they practiced as consultants or as employees, could not be held ethically accountable unless they all belonged to an engineering society which embraced and enforced its code of ethics. Employed engineers might be freer to practice if, as Whitelaw (10) suggested, they contracted for their services for definite periods and assignments. Such contractual arrangements were beginning to emerge in 1979, particularly for engineers in higher corporate echelons. If most engineers worked under contract, they could not then be laid-off wholesale, as were aeronautical engineers in the 1970 slump of the aircraft industry.

The reader may argue that individual contractual arrangements for employed engineers is too utopian. Yet all consultants execute contracts with the corporations or government agencies they serve. Employed engineers could likewise employ an agent to *bargain* for them *individually as professionals* rather than to allow labor unions to *bargain* for them *collectively as workers*. Aspects of this latter professional constraint will be treated later in Article 5E. One thing is certain: change will occur with time. It behooves us to improve societal conditions so as to do the most good for the greatest number.

Engineers, whether they serve as independent consultants or as corporate employees, can all practice their art so as always to hold paramount their obligation to the public welfare.

4

The Organization of
The Engineering Profession

4A. ENGINEERING PRACTICE vs. PROFESSIONAL UNITY

Early engineering practice was related almost exclusively to serving ancient city or nation states in military or civilian capacities. The engineers constructed public works like fortifications, roads, bridges, waterworks, buildings and temples. Their efforts, then, were not directed toward the production of consumer goods; artisans and craftsmen fulfilled that role until about 1750 A.D. when the Industrial Revolution began.

There was probably little thought given by these early engineers to unify themselves into a professional group, or by society to accord them professional status as it already had for the ministry, law and medicine. The engineering professionals of the early civilizations were far outnumbered by clergymen who nurtured souls, by physicians who healed bodies, or by lawyers who pleaded cases. The work of these three "organized" professionals was important in creating the socioculture and infrastructure needed for civilized society. These societal foundations formed the economic base that the "unorganized" engineers needed to produce the technological benefits which they developed. The service of all of these professionals was required to improve both the material standard of living and the civilized environment of society.

Dallaire (41) cites a similar concept Karl Marx (1818-1883) noted in DAS KAPITAL. Marx conceived of *society* as *consisting of two major parts:* 1) *basic technology* including tools, techniques, and all means of production and related services, and 2) *social organization* including everything else like business, government, law, religion, art, etc. Heilbroner felt that Marx over simplified the case when he (Marx) stated that, "The hand-mill gives you a society with the feudal lord; the steam-mill, society with the industrialist." Heilbroner felt that the steam-mill gives society the industrial manager. Managers are usually not capitalists; stockholders own industries.

The importance of engineers and scientists to society was, thus, evident to early philosophers even though the need for engineering professional unity did not become apparent until the 19th century. "The French social philosopher, Count Henri Saint-Simon (1760-1825) had an interesting way of measuring people's usefulness. 'We suppose,' he wrote, 'that France suddenly loses her 50 leading physicists,...chemists,.. physiologists,...mathematicians,...mechanics' and so on until 3000 savants, artists and artisans have been accounted for. What would be the result? A catastrophy, he said, that would rob France of her very soul...Now suppose, Saint-Simon continued, that instead of losing those few individuals, France was to be suddenly deprived of its entire "upper crust"--dukes, duchesses, ministers of state, judges and the 10,000 richest proprietors of the land. Result? The resulting loss, said Saint-Simon, would be purely sentimental." (41)

Thorstein Veblen, likewise, noted in his writings (56) that early 19th century industrial captains were builders of factories and machines, as well as shopmanagers and financiers. As time went on, corporate financiers assumed the dominant role, interjecting a conflict between the "work" of making goods and the "work" of making money. Financiers sought to maximize instant profit, not continual production, and thus created less output than was needed by society and increased unemployment (41).

DeLorean (57), an engineer, who as the youngest general manager of the Chevrolet Division and vice president of General Motors Corporation, also touches on this difference between engineering and administrative management. He cited examples of how the former was not unified enough to persuade the latter to accept what his engineering staff considered were sound technological decisions. Corporate officers delayed starting schedules for new models and failed to approve development of a light weight, fuel efficient automobile in the early 1960's. DeLorean claims the stockholders lost hundreds of millions of dollars in profits and the nation lost the small car world market to foreign competition in the 1970's.

62

We have already seen how (Article 1A) failure of the engineering profession and of industry to regulate themselves adequately led ultimately to the formulation of innumerable governmental regulatory agencies which are equally effective in curtailing production and profit. Of course, some of the regulations were necessary in an industrial society as complex as ours, but many could have been avoided if engineers had been unified enough to insist on professional accountability of all practitioners. Similarly, corporate managers must be held accountable for their economic decisions and not expect government bail-outs if bankruptcy threatens. They might favor more long term production policies instead of "quick-fix" short term solutions if engineers were unified enough for their suggestions to be accepted more often.

4B. SPECIALIZATION

No mention was made in Article 4A about the professional specialization which began in engineering soon after its *non-military* segment formed. Article 3F did cite the fact that this *civil* segment began to splinter into such specialties as mining, mechanical, electrical and chemical engineering because industry was becoming ever more complex technologically.

Before the splintering began, engineers were generalists, but the physical sciences which then formed the basis for engineering were simple enough to afford that luxury. As generalists, they began to unify professionally by organizing engineering societies to exchange technical information and to provide fellowship. In fact, the first societies were essentially supper clubs. The British Institute of Civil Engineers was one of the first (1818). The Royal Society, founded in 1660 in England, was likewise the outgrowth of informal weekly meetings begun in 1645 by British scientists.

Layton (58) notes in his REVOLT OF THE ENGINEERS that "engineering professionalism first appeared in America as an offshoot of the scientific variety" also. From 1829, when the Franklin Institute was founded in Philadelphia, to 1836 this society attempted to "gain patronage and prestige for science by linking it to practical activities, such as the investigations of steam boiler explosions." Its Journal was refashioned into three divisions, one of which was for *civil* engineering.

63

The first engineering society formed in the United States was founded in Baltimore, in 1839. It was followed by others in Boston (1848), New York (1852), St. Louis (1868) and Chicago (1869). All of these represented the whole nonmilitary, or civilian, spectrum of engineers and so were not splintered into specialties. Only the New York society had national aspirations as its name, The American Society of Civil Engineers, suggested. It lapsed into a moribund state after 1852, but was revived in 1867 and claimed to represent all American engineers not in military service. It also attempted to separate business from engineering, to restrict membership to professional engineers, and to stand for "the ideal of engineering as an independent profession" (58).

Mining engineers at that time challenged these ideals. In 1871, they founded the American Institute of Mining Engineers whose "fundamental aim was to serve the mining and metals industries" and which "did not restrict its membership to professional engineers...and showed little or no interest in professionalism" (58).

Thus began the splintering of the engineering profession. An initial, not all-exclusive, chronology is listed in Table 4A.

TABLE 4A - FOUNDING OF ENGINEERING SOCIETIES

1852-ASCE   American Society of Civil Engineers
1871-AIME   American Institute of Mining
            Engineers
1880-ASME   American Society of Mechanical
            Engineers
1884-AIEE   American Institute of Electrical
            Engineers
1888-ASNE   American Society of Naval Engineers
1893-ASEE   American Society for Engineering
            Education
1894-ASHVE American Society of Heating and
            Ventilating Engineers
1898-ACS   American Ceramic Society
1899-AREA   American Railway Engineering
            Association
1904-ASAE   American Society of Automotive
            Engineers
1908-AIChE American Institute of Chemical
            Engineers
1912-IRE   Institute of Radio Engineers

64

There are several hundred such specialty associations in engineering and related areas, including scientific and trade associations.

It is unfortunate professionally for engineering that so much splintering--or shattering--had to occur. This lack of unity, as well as the Grecian custom of considering any activity except contemplation as degrading, has stunted engineering's development as a recognized profession. This shattering has affected engineering education by eliminating the study of some basic engineering sciences like thermodynamics, electrical field and circuit theory, etc. from a few four-year curricula. Whether such graduates should be considered as engineers is a vigorously debated topic. There is a danger now, as Naisbitt (178) points out in MEGATRENDS, that educational programs "are moving from (those of) the specialist who is soon obsolete to (those of) the generalist who can adapt."

It would appear that all engineers should understand energy in all of its forms; i.e., chemical, electrical, mechanical, nuclear and thermal. Engineers should also understand the atomic structure of all materials like metals, minerals, plastics and composites, and of how these react when subjected to energy applied either slowly or with sudden impact.

The arguments for and against inclusion of a generalized core of basic and engineering sciences, and of lengthening the curriculum beyond four years for the first designated engineering degree, will be covered in detail in Chapter 6. The need for the study of professionalism, ethics, culture and the history of technology will also be cited then, particularly as all relate to management and leadership.

Whether or not the fragmentation of engineering into ever more specialties is advantageous, it is obvious that knowledge is increasing, as Figure 2A indicated. Enough basic and engineering science must be assimilated by all engineering practitioners. But, how much is enough? If some subject like thermal science is omitted from an "engineering" curriculum, should the graduate then be considered more of a technician or technologist? Or, if other cultural or management studies are also omitted, can the graduate be expected to provide any societal leadership?

The present engineering educational/apprenticeship model has produced about 50 percent of the managers for American industry. The production which that corporate model generated was the envy of the world, until it was challenged by European and Japanese conglomerates in the late 1970's.

Perhaps a comparison of the medical educational/ internship model might be worth considering to improve the dilemma facing the engineering profession. American physicians in 1900 required only three years of collegiate professional study with a one year internship to practice. Engineering then suggested four years of college. It still specified the same four years in 1980 for the first designated degree (see Figure 6E). Meanwhile, medicine found it essential to increase its collegiate study to include four years of pre-professional study plus four years of a professional program and a year's internship for general practice. Physicians aspiring to practice as specialists were obligated to an additional two or three years of study and residency in a hospital.

There has been at least as much knowledge generated in engineering as in medicine, and now the two disciplines are merging into such specialties as biomedical engineering and genetic engineering. Meanwhile, some engineering graduates have prolonged their formal study voluntarily so that perhaps 45% now earn an M.S. degree and about 7% a doctorate as is noted in Article 6F.

The great difference in these two educational models is that the medical one emphasizes general cultural and preprofessional study, followed by general study of the whole human body, before voluntary specialization is attempted after a required 9 year program is completed. Conversely, engineering still mixes some cultural general preprofessional and specialized professional study into a required 4 year program, and caps this with one to four years of specialized voluntary study. Admittedly, there is a growing need for professional specialists as civilized society becomes more complex technologically. However, there is an even more dire need for professional generalists to manage and lead their peers and to counsel legislators in the selection of proper long range goals.

The question facing the engineering profession is how the needed specialists and generalists can best be educated and unified for eminent practice, for industrial management and for societal leadership.

4C. TYPES OF ENGINEERING SOCIETIES

The splintering of engineering into such major specialties as civil, mining, mechanical, electrical and chemical engineering begun in 1871 has continued unabated. The 1980 Report of the Accreditation Board for Engineering and Technology (ABET) lists 21 program areas it recognized in engineering and 48 in technology. The ABET is represented by 19 engineering societies on its governing board. Details of this data are included in Appendix A.

The five major specialties noted above formed the basis for the five founder engineering societies that were organized from 1852 to 1908, and which are listed among others in Table 4A. All of those listed were essentially technical societies whose primary objective was to advance "the science and profession of engineering to enhance the welfare of mankind."* This they do by publishing technical journals and magazines, and by holding meetings at which the papers of these publications are presented and discussed. The ASCE, for instance, publishes 17 such monthly or quarterly technical journals and one monthly magazine which is also devoted mainly to technical information. A small part of the magazine is also devoted to engineering news items and more strictly professional matters. ASCE also publishes one journal quarterly devoted exclusively to professional matters. Other founder societies follow similar patterns. All of the other societies or associations are organized essentially to disseminate technical or business information.

One engineering society was founded in 1934 to consider only professional matters and unity. It is the National Society of Professional Engineers. NSPE membership is restricted to registered engineers licensed to practice in the several states and territories of the United States. Although almost 400,000 of the 1,000,000 individuals which the 1980 census lists as engineers were registered, the NSPE membership embraced only about 80,000 of these. However, the NSPE does not fare much worse than the founder societies,

---

*Quoted from Article I, Section 3, ASCE Constitution.

for their membership probably fails to include more than about one-third of the total in their specialty. It has been estimated that one-half of all engineers belong to no engineering society at all, but many hold dual membership in several.

Unlike the technical societies which disseminate the ever expanding, technical information, the NSPE restricts its activities to such professional matters as the employment practices and the economic status of engineers, engineering education, governmental legislation, industrial relations, public safety, professional development, licensing and ethics.

Three other engineering organizations should be mentioned. One is the American Society for Engineering Education (ASEE) whose membership is open to educators interested in the cultural, technical and professional education of engineers. The other two restrict their membership by appointment only. They are the National Council of Engineering Examiners (NCEE) composed of representatives of state licensing boards, and the Accreditation Board for Engineering and Technology (ABET) composed of representatives of the engineering societies listed in Appendix A.

4D. ORGANIZATION AND UNITY EFFORTS OF SOCIETIES

Little thought was given originally to planning the organization of engineering as a profession. As was noted in Article 4C, the several technical, professional and educational societies did develop. Their functions frequently overlap. Their functional objectives have been diagrammed in Figure 4A which shows how the societies were organized into the latest unity effort in 1980 under the American Association of Engineering Societies (AAES) and its four councils. Two of the twenty-four societies which were represented on the AAES Board of Governors also played a major independent role in fulfilling the objectives of two of the councils. The ABET was to carry on engineering college accreditation and other educational functions associated with the Educational Affairs Council for Professional Development (ECPD). Similarly, the NSPE was to fulfill an independent, non-technical, professional function associated with the Professional Affairs Council since these coincided closely with those of its own objectives enumerated in Article 4C.

68

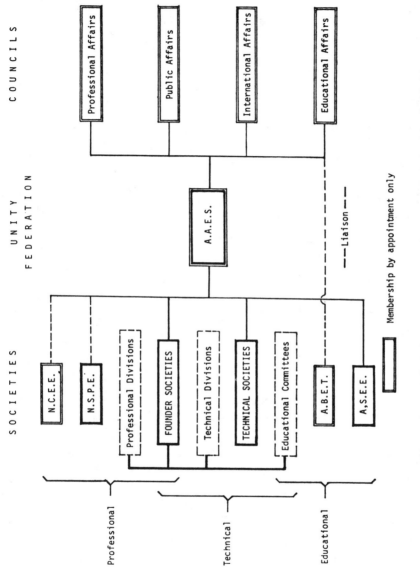

FIGURE 4A - AAES ORGANIZATION OF THE ENGINEERING PROFESSION 1980-83

Originally NSPE was one of AAES's 43 member societies. However, NSPE and 17 other members withdrew according to the September 1983 report of the AAES Trien-
nium Review Committee (187). It recommended sweeping
changes in organization and focus and eliminated the
Educational Affairs and the Professional Affairs Councils. The Committee noted that AAES had "served as a
broad based forum for engineering affairs," had "provided a mechanism to address technical issues of importance," and had "established communication with various
publics." Nevertheless, the Committee noted that AAES
has "not effectively brought unity of action...has not
effectively dealt with some key issues, has not managed
its affairs prudently, and has not effectively used the
potential resources available to it through its member
societies" (187). The Committee made a number of recommendations besides eliminating 2 of its 4 Councils.
These reduced and revised the role of the AAES president from one of a spokesman for the engineering community to one with primary emphasis on AAES administration.

The AAES was not the first attempt to unify engineers as individuals or of the societies to which they
belong. In fact, there were many attempts as Figure 4B
shows (60). These data, abstracted from Layton's book
(58), show the first effort to have been that of the
Franklin Institute in 1829 which was described previously in Article 4B. The next major effort was that of
ASCE in 1852 which endeavored to unite all civil, i.e.,
non-military engineers into participatory membership.
That attempt was ended in 1871 when AIME began the process of forming specialty organizations.

The ASCE attempt was organized so that all individual members could elect all officers of its Local
Branches, State Sections and the National Board of Direction. Service on all technical or professional divisions and committees is voluntary, and by appointment
at least approved of in principle by the Board. It
was, and still is, a democratic organization whose governing officers are elected and not appointed.

The American Association of Engineers (AAE),
founded in 1917, also based its unity effort upon a
participatory membership of individuals. The AAE
quickly grew to one of the largest societies with over
20,000 members within four years. It failed soon after
because its members were preoccupied with their own material interests rather than professional objectives.

70

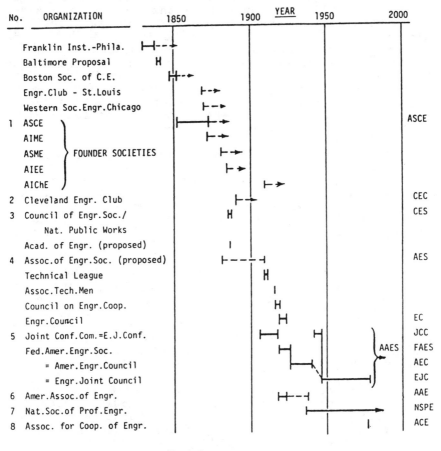

No.	ORGANIZATION	YEAR	
		1850 1900 1950 2000	

Franklin Inst.-Phila. ASCE
Baltimore Proposal
Boston Soc. of C.E.
Engr.Club - St.Louis
Western Soc.Engr.Chicago
1 ASCE ⎫
AIME ⎪
ASME ⎬ FOUNDER SOCIETIES
AIEE ⎪
AIChE ⎭
2 Cleveland Engr. Club — CEC
3 Council of Engr.Soc./ — CES
  Nat. Public Works
  Acad. of Engr. (proposed)
4 Assoc.of Engr.Soc. (proposed) — AES
  Technical League
  Assoc.Tech.Men
  Council on Engr.Coop.
  Engr.Council — EC
5 Joint Conf.Com.=E.J.Conf. — JCC
  Fed.Amer.Engr.Soc. — FAES
    = Amer.Engr.Council — AEC
    = Engr.Joint Council — EJC
6 Amer.Assoc.of Engr. — AAE
7 Nat.Soc.of Prof.Engr. — NSPE
8 Assoc. for Coop. of Engr. — ACE

AAES

Unity Effort ⊢—⊣ Started ⊢—⊣ Cont'd. ⊢ — →
                Stopped

Data abstracted from "The Revolt of the Engineers"
by Edwin T. Layton, Jr.

FIGURE 4B - ENGINEERING UNITY EFFORTS (60)

One other effort to unite all engineers was begun by NSPE in 1934 but, as already noted, it restricted its participatory membership only to registered engineers. The NSPE endeavored in the 1960's to broaden its admission standards by admitting non-registered engineers for five years if they held the highest grade of membership in any of the founder societies, and if these founder societies would thereafter require registration for such highest grade of membership as ASCE does. Unfortunately, the NSPE membership failed by four percent to garner the necessary two-thirds needed to amend its constitution after most of the founder society boards had approved the concept.

All other attempts to organize a national umbrella organization have failed to provide participatory membership of individual engineers, and were essentially confederations whose board members were appointed, not elected. A number of these are listed as items No. 3, 4 and 5 of Figure 4B, the last of which, the Engineers Joint Council, was replaced by the AAES on January 1, 1980.

Unfortunately, this series of confederations of societies aroused scant interest among engineers, perhaps because they had no direct opportunity to elect their council representatives, nor to pay dues directly. Instead, council members of the unity association have usually been appointed by a society's president and approved by its board of direction; they were twice removed from their membership.

A less formal but more effective effort to unify engineering society cooperation on public issues was initiated by NSPE in 1971 through Intersociety Liaison Agreements. These joint, voluntary endeavors were restricted only to issues which interested one or more societies enough to cooperate. By 1983, 25 societies had joined.

A more formal unity organization employing the best feature of the AAES with its four councils, and of the NSPE with its Intersociety Liaison Agreements, was described in 1983 in a futuristic, conjectural history of engineering unification looking backward from the year 2020 with 20/20 hindsight.

"It was conceived as a global, participatory one using national grass-roots society membership wherever that was available.

Initially, this International Alliance of Ingeniors* (IAI) bypassed organizing on a national level for two reasons: a) industrial
and ingenioring worldwide activity had already penetrated national political boundaries as though they were sieves, and b) previous unification attempts of society *confederations* had generated too much jealousy...This
new global *participatory* society restricted
its membership...to include only mature, bona
fide ingeniors. No corporate or university
memberships were permitted.

There was one other reason for organizing IAI on an international basis. That was
related to the failure of most national engineering societies to integrate the professional activities of two distinct types of
ingeniors; i.e., those who design, manufacture and service products, and those who design, build and operate the private and public facilities on which the industrial and
societal infrastructure rests...IAI was successful in uniting product and facility ingeniors...by channeling their ability to
serve society into four divisions. These
were the divisions of technology, education,
government and employment" shown in Figure 4C
(181).

A different type of national unity organization
was also proposed in 1974 which paralleled the AAES
structure, but differed from it by providing for membership of individual engineers as well as of societies. "Figure 4D represents such a plan. It shows
various national technical and professional societies
with their regional, state and local units grouped under a national unity organization that might be called
the "United Engineering Societies" (UES). "Dual membership in this and the engineer's technical or professional society would be required but restricted in the
unity organization only to 'graduate' engineers or
their equivalents. A bicameral legislature could provide separate representation for the constituent societies and for the individual members" (22,60,62,116).

---

*See footnote on page 46 for the derivation of
ingenior.

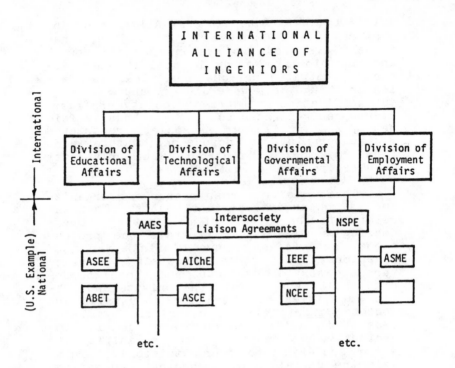

FIGURE 4C - IAI ORGANIZATION (181)

It also provides for an international global association, the United Ingenieur Diplomates (UID), but
only of those who have been elevated to diplomate status.

"The Ingenieur Diplomate* citation
should be awarded only to those who apply and
then only if their total technical education
ensured its currency, if they had achieved
specialty status by having passed national
(specialty) board examinations, and if their
leadership in influencing societal technolog-

*The foreign spelling of engineer; i.e. ingenieur, is
preferred because of its worldwide usage. See also the
title of ingenior suggested in Article 3A.

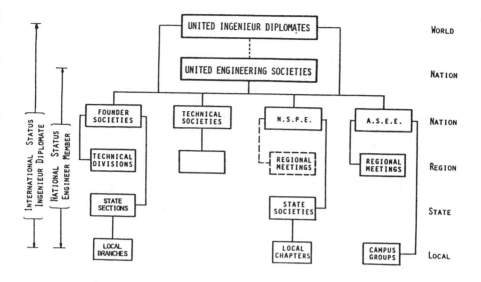

FIGURE 4D - ENGINEERING UNITY (60,62)

ical priorities was evident. This proposal
calls for issuance of the Diplomate citation
by the applicant's own technical society of
any national origin, under policies established jointly by all societies. Lapse of
any Diplomate's leadership role could relegate him to emeritus status...

If the engineering profession would in-
augurate specialty boards in its several divisions, the public would also benefit
through ever improved consumer goods and services, and the engineers involved would have
a professional incentive to continue their
development. Examinations for such specialty
boards could easily be provided by each major
engineering technical society, with the general requirements specified by the Accreditation Board for Engineering and Technology
(ABET).

75

The American medical profession has for years found specialty board practice advantageous for both the public and the physician...

After...national democratic unity is achieved by organizing the UNITED ENGINEERING SOCIETIES, and national specialty boards are provided by the national technical societies, arrangements should be made to confer diplomate status on all who apply and who have also demonstrated professional leadership. These diplomates of all nations could then unite to form a global organization--the UNITED INGENIEUR DIPLOMATES--dedicated to "hold paramount the health, safety and welfare of the public" as the engineering profession carries on its public purpose.

It is unlikely that much can be done to enhance professional engineering service, dedicated to societal progress by always enhancing the public health and safety, without first planning for the profession's global unity. Failure to do so only provides all governments with more reason to enlarge their regulatory bureaucracies. But the growth of governmental agencies is continually accompanied by more constraints on professional practice and, in fact, on all liberties, personal and political as well.

The growth of this collectivized government can only be stopped and then reversed by having all professionals assume complete responsibility for their practice. Engineering unity will be needed to first convince the public and the legislatures that it is in the economic and political interest of all to 'let the buck stop' with the engineer-in-responsible-charge...That kind of practice has served society and western civilization well for centuries" (22).

That kind of practice began with the Hammurabi building code of about 2000 B.C. which required the death penalty for any builder whose structure collapsed with loss of life. Its message is crystal clear. It did not stifle innovation. It will be discussed later in Article 5E and is illustrated in Figure 5C.

One of the best descriptions of "the long quest for unity" of international engineering relationships "as a structured entity" is included in Wisely's book THE AMERICAN CIVIL ENGINEER (63). He describes domestic and global efforts that began in 1874 and have continued to the present. Table 4B lists some of these efforts.

TABLE 4B - INTERNATIONAL ENGINEERING UNIFICATION

1874-AEA	Association of Engineers and Architects of Austria
1889-IEC	International Engineering Congresses begun in Paris and held also in Chicago-1893, Paris-1900, St. Louis-1904 and San Francisco-1915
1921-ICE	Institute of Civil Engineers (Great Britain) General Engineering Conferences-Paris
1929-WEC	World Engineering Conference (Tokyo)
1939-BAEC	British-American Engineering Conference
1946-WEC	World Engineering Conference (Paris)
1949-EUSEC	Conference of Engineering Societies of Western Europe and the U.S.A.
1951-UPADI	Union Panamericana des Associaciones des Ingenieros
1968-WFEO	World Federation of Engineering Organizations

These unity efforts considered such functions as education, standards, manpower, membership exchange services, international registration, etc. Individual societies served as catalysts.

"Best results" were "realized with the Joint Conference Committee approach...personified at the levels of the presidency and chief executive officer"...who served "only to define the problem and decide on strategy" (63).

Frequently, the work was then assigned to national society committees. But this effort was frequently inadequate because the members may have had other time-consuming duties and committee assignments.

"The greatest value of the tremendous brainpower potential of the membership resource has not been utilized effectively... and this avenue for professional service deserves study. This problem-solving capability could best be applied to the problems of society rather than those of "the Society." New means for mobilizing this brainpower for the guidance of political and public policy decision-makers must be developed" (63).

The reader can now see that desire or the need for professional unity is not new. Realizing it will not be easy. However, if it is to be effective in the future, it may be that much more participatory activity will be needed to accomplish the tasks and to interest the membership. The availability of current instant electronic communication, and the ability of the unity organization to conduct statistically significant membership polls should provide for easy formulation of technological policies on public issues.

Engineers seldom work for themselves anymore on small local projects. Rather, the majority of engineering graduates work for large industrial firms or governmental agencies where their work may have global or even extra-terrestial implications. Because of the nature of the role between employer and employee, the average engineer has little impact on the socio-political aspects of the job he performs, yet he is uniquely educated to contribute logically to the formulation of technological issues. Such issues are usually the purview of the employer who has business or legal training and who may not be the best qualified person to speak on the implications of the technological decision process which he controls.

Often, the only way an engineer can have a participatory role in the decision process is either to become involved in management or politics, or to speak through his professional society(s). Only a limited number can have an effective voice by turning to management or politics. Further, the engineer who turns in this direction may no longer necessarily project an engineering point of view (138).

78

Thus, for all practical purposes, the various professional engineering societies represent the mode by which the majority of engineers have a voice on the national or international socio-political implications of engineering technology. The effectiveness of this process is muted by the fragmentation or size of engineering societies in general, especially those which claim to speak for the professional interests of engineers. Further, many such societies are much too narrow and, if not, rarely take stands on political issues outside their own narrow and often vested interest. Yet the need for technology-related societal leadership is so urgent today. It will be discussed in more detail in Chapter 8.

4E. QUALIFICATION FOR PRACTICE

Two pertinent questions surface when society addresses the questions of 1) how to qualify any professional for practice, and 2) whether to even bother to do so. Heated arguments have embraced both questions. The qualification procedure necessarily involves the total educational process and of its evaluation for the novices who pursue it. The requisite qualifications can be acquired by apprenticeship, education or experience. Time has proven all three are needed for optimum professional service.

Originally, the apprenticeship system was the only means available to educate specialists in any craft, vocation or profession. The early craftsmen who supervised the novice's training were experts in their calling as well as managers of businesses or of governmental enterprises. They had established their reputation by direct competition. Those supervisors who could not earn a living were unable to afford an apprentice, let along attract any. The masters exchanged their proven competence and ability to transmit it to novices for the inexpensive labor the novices supplied. The masters could ill afford to certify incompetents for long. The question of competence was determined in the market place, or in governmental service.

The clergy were the first to break this tradition when they established schools for priests in Egypt about 3000 B.C. This formal education proved to be more efficient for the transfer of, and the distillation of, the ever growing body of pertinent knowledge. It is likely that these early professional schools retained some practice-oriented apprenticeship training, but may not have shortened the total period of study.

The Greek academies, as we shall see later in Chapter 6, emphasized educational fundamentals but ignored vocational studies, except perhaps in medicine as noted in Article 3B. Early Roman education was more practical for it included basic studies initially and later emphasized "citizenship; i.e., professional education like military training for some, and oratory and management of public and private affairs for others. Both educational programs were ultimately combined, but by then their importance deteriorated as the Roman Empire declined and barbarians overran Europe" (64).

Schools for physicians and lawyers, founded in the 11th century, will be described in detail in Chapter 6. All such educational programs were developed to qualify novices for practice and to certify their competence. But this raises the question as to who should teach the novices, and who should certify them. In the apprenticeship system the experienced master craftsman did both, but his own economic survival was intimately and swiftly related to this educational effort. In current engineering academic systems the tenured faculty do both, but they are virtually immune from the judgement of the market place and sometimes lack significant practical experience (27).

The major questions can thus be reduced to who should determine qualifications for professional practice and certify competence. Business and government accept screening of applicants by academic certification, yet in most cases the engineering baccalaureate signifies the state of the student's technical literacy rather than his ability to practice. The market place will ultimately determine his competence and the reputation of his college, but the time-lag is usually too long for the evaluation to be significant.

Who else should certify competence? Society has provided governmental agencies to do so for a spectrum of occupations from licensed professions to certified vocations. Society has also accepted certification by professional societies, but these carry no legal significance. If governments or societies are to certify an individual for a profession, they can do so by *examination* of his educational program and/or *evaluation* of his experience. Both should be checked, for the latter cannot be examined. Qualifications of all professionals and specialists are specified, but not certified, by the managers of industries and governmental bureaus who employ them.

Certification today is granted for life even though all practitioners know that engineering practice is soon outmoded as knowledge expands, and as new techniques and materials are introduced. Some state licensing boards are now reexamining all licensees every five years. This guarantees the current status of all practitioners, and precludes the exclusion of new applicants by continually increasing the rigor of the examinations.

Society has thus provided the three routes of apprenticeship, schooling and experience by which novices can assimilate the required minimum knowledge to practice a profession. In today's complex technological world, all three avenues must be pursued if the engineering professional's ability and willingness to protect the public's health and safety are to be preserved at all times.

Society has established two ways to protect the public from injury due to faulty consumer products or technical systems. Either it holds the manufacturer financially and criminally responsible for the products he sells, regardless of whom he employs to make them; or it licenses the professional to practice and holds him legally responsible for his advice and service to the public. The licensing system originated because the public is seldom knowledgeable enough to determine whether any professional is qualified to practice. Licensure, however, merely guarantees that a professional has met the required minimum criteria, but licensure cannot guarantee that he will not make mistakes.

How, then, could the public be better protected? Two ways might be suggested: 1) eliminate all restraints as was suggested at the beginning of Article 4E; 2) restrict professional practice only to those who volunteer to be held accountable for their service and to submit to disciplinary action by their peers if they violate a code of ethics.

The first suggestion would relegate professional practice to the rules of the market place; i.e., caveat emptor--let the buyer beware. The buyer may be aware, but is not competent to evaluate most professional's qualifications. The buyer could, though, determine how the professional acquired his competence, and whether his clients recommend his services. For instance, a patient might inquire as to where a physician studied medicine and completed his internship, as well as whether he had passed national specialty board requirements for some medical specialty. The patient might also inquire whether the physician was insured for possible malpractice. Thus, people could examine the professional's credentials, and reputation, and his ability to secure insurance for his practice, to judge his qualifications.

Similarly, the client of an engineering consultant, or customer of a corporation, interested in purchasing some device or system, could determine whether any financial loss due to its inability to function properly was covered by insurance. No insurance company would knowingly insure a physician or engineer without first evaluating his ability. It is likely that the company's examination might be more rigorous than state licensing criteria, for it stands to lose money if its judgement is bad. The insurance company, thus, could police professional competency. However, the argument could be made that, given sufficient volume of insurance business, the insurance company would merely charge all policy holders sufficient premiums to cover losses and underwriting expenses.

It is also possible for the buyer of consumer products to determine whether his intended purchase has been certified by some established testing laboratory or engineering society. Two such illustrations come to mind: the certification of the Underwriter's Laboratory for electrical and mechanical products, and the American Society of Mechanical Engineers for high pressure vessels. Both attach labels or seals guaranteeing approved manufacturing techniques and materials. No insurance company will insure any high pressure vessel unless it bears the ASME seal.

It is unlikely that the free market system of unlicensed professional services (Suggestion No. 1) would ever be accepted, but it must be admitted that it does have some advantages. Perhaps, though the average customer or client would rather avoid the responsibility of selecting an acceptable professional, and let some recognized agency do so for him. He then has someone else to blame--and sue--for any error in his own judgement.

The other suggestion (No. 2) would require some way to first identify the professional and then to require that he be bound by an ethical code enforced by a recognized professional society. Of course, the public then would have to patronize only those professionals who were properly certified and accountable. Ethical accountability will be covered in more detail in Article 4G and Chapter 7.

"The importance of specifying proper qualifications for professional practice becomes even more important if these qualifications are to be broadened to include societal leadership as well. Now none are. The imperative need for technological leadership is finally being recognized as a vital element of democratic governance. Civilized society needs this added input if its governing bodies are to establish appropriate technical goals for posterity. The world energy crisis of the 1970's illustrates the need; the recent creation of the Office of Technology Assessment in the United States illustrates one effort to fill that need.

The point to emphasize here, however, is that neither the engineering profession nor any other profession includes leadership criteria now in the specific qualifications for professional practice. The following chapters will attempt to show that these criteria will have to be added to qualify future novices for technological leadership as well as practice, and that the added criteria will also have to be absorbed by a significant percentage of qualified practitioners as well. Failure of the professions to do so will relegate the free enterprise system of Western Civilization to one of increasing bureaucratic control and decreasing individual freedom" (12).

The reader can see by now that no system will be foolproof. It is evident, however, that the better the technical and ethical criteria that are used to qualify the professional, the better will be the service rendered. The ways by which such criteria may be established need now to be addressed.

4F. METHODS OF QUALIFICATION*

Professionals may have their qualifications for practice evaluated and/or approved by one of the following four principal methods.

ITEM 1 - QUALIFICATION BY LICENSURE
License to practice is granted by state boards of professional and occupational regulation to any applicant who meets the minimum acceptable professional educational requirement, who has acquired sufficient practical experience acceptable to the board, and who passes a written examination.

ITEM 2 - QUALIFICATION BY REGISTRATION
Registration may require compulsory application of individuals to engage in some specified activity, and may list them in an official register, but may exclude the need for any specified education or demonstration of competence. (Example: lobbyist. Exception: nurse--a registered nurse (RN) is really licensed to practice.)

ITEM 3 - QUALIFICATION BY CERTIFICATION
Certification requires some recognized organization (governmental agency, church, professional society, college) to vouch for the specified skills or education of the individuals certified, as determined by examination and/or experience. (Example: CPA--singular notable example of legal certification.)

ITEM 4 - QUALIFICATION BY CREDENTIALIZATION (Non-legalistic)
Credentialization implies that proof of the validity of specified credentials has been verified by some recognized organization for a given individual. (Examples: college degree, professional society membership grade.)

"All methods employ examinations on related subject matter and/or evaluation of experience, usually under qualified professionals. The usual lengths of educational pro-

---

*Extracted from the NSPE Report noted in Reference 12.

grams and the subsequent internships vary for the several professions as Table 4C shows. In some states, demonstration of continuing competence is now being authorized for re-licensure. Documentation of suitable activity, like formal study, attendance at professional meetings and workshops, etc., and active practice is prescribed.

TABLE 4C - EDUCATIONAL PROGRAM AND
INTERNSHIP LENGTHS

| | Time in Years | | Type of |
	Education*	Experience	Qualification
Accounting	4	2	Certification
Architecture	6	4	Licensure
Engineering	4	4	Varies
Law	7	0	Licensure
Medicine	8	1	Licensure
Ministry	7	1	Ordination

*Includes pre-professional plus professional study.

There may or may not be legal penalties for practicing a vocation as an unregistered individual. There are such restrictions on lobbyists; thus, their registration might be considered as a license. Professional engineers must be licensed to practice if they offer their services to the public, but are usually referred to as being registered. Actually, they, like physicians and lawyers, are registered and licensed. However, most engineers employed by industry and government are exempted from such legal registration" (12).

Some governmental agencies, however, like the Bureau of Reclamation and Corps of (Army) Engineers, as well as many corporations require that their project engineers-in-charge be legally registered.

"Certified Public Accountants (CPA) are legally certified after they serve a two-year internship and pass a state board examination. Non-certified accountants are not denied the legal right to perform accounting

85

functions but their documentation may not be acceptable legally.

Ministers are ordained to conduct church services by their church heirarchy. Clergy with a pastoral assignment in a state are authorized to validate marriage licenses, as are Justices of the Peace or Judges.

Medical specialists, like surgeons, pediatricians, etc., are so certified by one of their professional societies as having met their specialty board requirements for additional study, experience and examination. However, they may practice any phase of medicine licensed only as an MD. Unless they qualify, they cannot list themselves as specialists without being disciplined by their peers for an ethical violation. The American Academy of Environmental Engineers (AAEE) likewise issues a certification as a diplomate to a specialist who has met specified educational and experience requirements and passed an examination. However, a licensed environmental engineer need not be a diplomate to practice" (12).

The discussion of which method would be best to qualify engineering professionals for eminent practice and societal leadership would, if assembled, fill volumes. I am inclined to favor licensing of all practitioners to 1) pinpoint direct legal responsibility, 2) to provide a mechanism to enforce ethical accountability, and 3) to serve as the foundation of professional unity and societal leadership. The public would then know "where to stop the buck" and might amplify its implicit faith in engineering. The only other alternative to the three objectives just cited would be required certification by a single engineering unity organization.

The unprecedented success of the moon landings and the space shuttle flight makes it hard to convince corporate management, governmental leaders and the public that engineering is not perfect, and that its professional service could be improved. Industry is, as a whole, content with the present situation, even though a few firms have paid huge sums in damages and legal fees for cases related to waste disposal, structural failures and faulty products (53,54,55). Engineering's

86

accomplishments have been brilliant worldwide, but its technological leadership has been lacking. If the status quo continues, industries will have to be content with increasing regulation, society will fail to benefit from economical long-range technical missions, and "engineers will have to remain just a bolt in the huge system."

If the need for, and the methods of, qualification to practice as an engineering professional were changed as suggested, industries, nations, people and practitioners all might benefit.

4G. PROFESSIONAL OMBUDSMEN

It was Wickenden (66) who, as early as the 1920's noted that "...a profession must guarantee to the public the trustworthiness of its practitioners. In return, the public protects the profession from incompetent judgement of laymen..." and grants it "self governing priveleges...Professional status is...an implied contract to serve society, over and above all specific duty to client or employer..." Wickenden continually advocated the need for The Second Mile of effort, and believed that remedying the engineering's shortcomings could not be achieved either by "keeping the boys longer in college," or by restricting the engineering title only to those licensed.

Perhaps the principal shortcoming of engineering so far has been its failure to provide a mechanism suitable for the neutral review of proposed technological industrial operations or designs allegedly threatening the public safety. Occasions have arisen when the engineers involved were concerned enough to protest when their technical decisions were overridden by non-technical administrative management, and when public safety and welfare were involved. Codes of ethics require the engineer to report such incidents "to the proper authority," but that authority has never been defined.   It could conceivably be a governmental bureaucracy, an engineering society, or his employer's corporate management.

An attempt was made in the 1970's to define this ambiguous authority as an engineering society's office of professional ombudsman. Such services are available now on a confidential basis to engineers, industries and governments in need of help on societal issues that could involve practices endangering the public safety

87

or wasting its resources. The ombudsman's findings would be revealed only if the product or system were unsafe and its scheduled production or operation were imminent. Publication of the investigation in such cases might have no coercive legal power, but the report's ability to sway public opinion could be substantial if the public were aware of the profession's dedication.

Several societies now have such ombudsman's offices. Among these are the ASCE Confidential Ethics Advisory Service and the NSPE Public Advocacy Review Board. The American Chemical Society has a similar service. The By Laws of the Institute of Electrical and Electronic Engineers (IEEE, which resulted when the AIEE and IRE merged) specify, in Section 112, Paragraph 4, that assistance will be available as follows:

> The IEEE may offer support to any member involved in a matter of ethical principle which stems in whole or in part from such member's adherence to the Code of Ethics, and which can jeopardize that member's livelihood, compromise the discharge of such member's professional responsibilities, or which can be detrimental to the interests of IEEE or of the engineering profession. All requests for support...shall be submitted initially to the Member Conduct Committee...IEEE support of members requesting intervention or amicus curiae participation in legal proceedings shall be limited to issues of ethical principle...

> The Board of Directors, or the Executive Committee upon approval by the Board of Directors, may publish findings, opinions or comments in support of the member, and take such further action as may be in the interests of the member, the IEEE, or the engineering profession.

The ombudsman mechanism described above offers peer-group support to engineering professionals who are fulfilling their ethical responsibility to "hold paramount the safety, health and welfare of the public in the performance of their professional duties" (See Appendix B, Fundamental Canon 1, Code of Ethics). Such support is also available in a few large industries and

88

governmental agencies which have "in-house" safety divisions responsible directly to the chief executive officer to which employees can report possible safety hazards without fear of reprisal. Too few industries or agencies have such safety divisions. Whenever they are unavailable, the engineer can always appeal to established engineering society ombudsmen.

The mechanism was developed because ethical professionals did occasionally face shortsighted managers more interested in profit than safety, or were involved in technical situations like nuclear power in which there was a decided difference of opinion (69). A few of these are cited here merely to illustrate the types. Some of these and others are described in CONTROLLING TECHNOLOGY: ETHICS AND THE RESPONSIBLE ENGINEER by Stephen Unger (185).

The San Francisco Bay Area Rapid Transit (BART) network malfunctions, the DC-10 rear cargo door failure and the engine maintenance procedures, the Lockheed C-5A cost overruns, and the Goodrich aircraft brake are examples (53,54,55,67). These four cases alone cost industry and taxpayers hundreds of millions of dollars and 672 lives. Similarly, the disposal of industrial wastes like kepone, PCB, pesticides and other toxic chemicals have permanently poisoned burial sites that resulted in staggering clean-up costs, legal fees and damage claims. If the engineers involved in such practices knew that they--not their employer--would be held accountable for burying industrial wastes in land fills they would explore alternate safer solutions. Actually, in the case of the Love Canal (68) the toxic wastes were buried in the 1940's using procedures that meet 1980 Environmental Protection Agency (EPA) standards. Subsequent unsafe construction on these land fills was authorized by a local government agency.

One last illustration will be cited to emphasize the need for professional ombudsmen. The aid of engineering societies proved helpful to the nuclear power industry in November 1976 when energy referendums were held in six western states. TV spot announcements and public meetings were scheduled in which engineers endeavored to educate the public by presenting the advantages and disadvantages of nuclear power generation. In all six states the vote favored its continuation.

Professional societies should join forces with corporate engineering managers, encouraging employed

engineers to engage in those societal issues which are
linked to their employer's operation. Industrial leaders are keenly aware of their personal accountability.
The production of unsafe products or services places
them in double jeopardy: their corporate careers and
legal liability may be involved. Engineers involved
may likewise be held accountable.

The simplest way to protect the public's safety,
the industry's profits, and the engineer's career is to
solve issues locally where the professional's practice
exists. The use of the ombudsman's office may help
when disagreements cannot be resolved satisfactorily.
However, it should not be involved unless the public
would be endangered or its resources wasted. In either
case, team action--not legal action--would be beneficial for all concerned.

5

Economic and
Political Constraints

5A. FREE MARKET CAPITALISM VS. SOCIALISM

The preceding chapters addressed the purpose of and the need for professions in general, as well as the heritage and organization of the engineering profession in particular. The discussion so far endeavored to convey the idea that the practice as an engineering professional is constrained in a complex societal environment. If his emerging role is to serve society as an ingenious practitioner and as a collaborative leader, he must master his technical specialty as well as overcome the restrictions of the prevailing economic and political systems within which he operates. His constraints are both technical and political; technical because scientific knowledge and the art of engineering are continually expanding, and political because social customs and governmental leadership are always changing. Additional constraints will not guarantee a riskless society but risk assessment is always necessary (Article 2A, p. 18).

Nevertheless, the governmental constraints which are developed, in response to serious risk assessment can actually reduce unnecessary constraints. For instance, early development of criteria for the disposal of nuclear and chemical waste would have precluded the need to develop the restrictive directives, which later had to be superseded, as the magnitude of industrial pollution became apparent.

Unnecessary constraints will suppress industrial output as effectively as central planning stifles production in socialistic states. The absence of centralized directives in the complex free market economies, and the presence of simple personal incentive systems, are what enable capitalism to out produce state socialism by at least one-third. If the role of the engineer in both capitalistic and socialistic states is not only to protect the public's health and safety, but to maximize production, conserve material, decrease obsolescence, and minimize pollution, he should favor the most productive economic system. Statistics overwhelmingly favor the free enterprise system in this regard.

The period from 1950-70, for instance, witnessed average real incomes in the capitalist world doubling every 15 years with unemployment and inflation stabilized at about 2%, while real production in Communistic economies (Comecon) was doubling only every 20 years (70). Supposedly, there is no unemployment in Comecon, but those who have visited these countries note that although all adults there may be employed, not all work. Overmanning is commonplace. There may be no inflation but there are long queues. The misery index; i.e., the sum of inflation and unemployment indices, was about 4% in 1962, 8% in 1973 and 20% in 1981 or the same for both communist and capitalist governments. Market economies in 1980 were growing only half as fast as they did a decade before. "The planned economy is simply not as good at producing goods and services that people want...The production achieved from each extra unit of investment (in capitalistic economies) is about twice what it is in Comecon" (70).

There is however, a difference between the two systems. "The gap in efficiency is smallest in areas to which the Russians give highest priority and in which the West does not use a full market system. Although the West's economy is bigger, the other side puts a larger proportion of its gross national product into military spending" (70). These proportions were estimated in 1981 as follows: USSR - 13%; Warsaw pact 10%; USA 6%; NATO 4.2%. "The more completely capitalist the structure and attitude of a nation is, the more pacifist--and more prone it is to count the costs of war" (70). Moreover, the West's channeling of capital from corporate investment to welfare payments is continually weakening its relative strength.

It is, thus, apparent that engineering's productive ability is intimately involved in both economic systems, and that more consumer goods are available in capitalist economies. It is also clear that the world leadership in both East and West does not regard the common man's welfare as a first priority. Otherwise the leaders would not divert materials to build battleships like the Bismark and the Hood--or aircraft carriers--and sink them in the oceans. Nor would the leaders divert productive capital to non-deserving welfare recipients. It is sad to conclude that all of these decisions in government, or industry, or education or even in the church structure are made by the educated people. Educators can only wonder where they went wrong. They educated these leaders.

There is a need for change in leadership that understands useful productivity and technology and that will develop a political-economic system better organized to "harness politics for the use and convenience of man" (148). But, whether our present society will change for the better or not, it will not be a riskless one either technically, economically or politically.

5B. POLITICAL FREEDOM--PROTECTION OF LIFE, PROPERTY, SPEECH, PRIVACY

Risks are an inherent part of social progress. Those social structures which are the most flexible are best able to minimize risk and maximize production. The western democracies, with their mixed economic systems straddling the gap between free market capitalism and state socialism, are far more effective than totalitarian regimes in improving living standards, perhaps because of the greater political freedom the democracies guarantee.

It should be noted that although engineers can practice their art in either free or totalitarian systems, they might not be considered as professionals--or even of having a profession--unless their paramount loyalty is to the public and not to the state. Their other loyalties to their employer or client, to their peer group, or to their families are important but cannot be dominant if the public welfare is to be protected.

Those political freedoms involving the protection of life, property and even privacy are intimately related to engineering practice and the technical and economic risks it constantly embraces.

"In earlier times engineers were able to take greater risks because their products were simpler and populations were sparser. When accidents did happen, fewer people were involved. No engineer, however, would voluntarily design an "unsafe" product. Yet he realizes nothing is absolutely safe, and that

93

he must deal in probabilities related not
only to the production of his designs, but to
their use by the public. Whenever significant progress is made, he works beyond the
state of the art. He always tries to at
least make his creations "fail safe" but must
continually balance the costs of their manufacture and operation against their usable
life and durability. At times his technical
judgement is overruled by the collective wisdom of his peers. Very rarely has such re-
versal resulted in the production of unsafe
products or systems" (12).

A similar comparison was made by Judge Markey for
the way laws changed to protect the public in a democratic society. "In simpler times," he said, "the law
sought the maximum freedom of man to do whatever he
wished, so long as he did not hurt others. If others
were hurt, the law acted after the event...When man
moved on...the law moved on to prohibit what might hurt
others...Many recent laws now have begun to...prohibit
man from hurting himself" (71). If that trend were
continued, everything in life would be ordered or forbidden, and engineering practice would be hopelessly
constrained by regulations designed to generate a riskless society. Laws are inadequate to deal with risk
assessment and courts should be expected to rule only
on questions of law and to interpret policy--not to
evaluate science or technology. The people, acting
through their representatives, should make the basic
decisions controlling public health and safety by ex-
ploring the associated risks involved in any issue as
well as their overall economic impact, and by changing
the laws accordingly (177,180) as was suggested in Article 2A, p. 18.

That legislators must be assisted by investigative
staffs of knowledgeable scientists and engineers is
self evident. "Engineers alone cannot determine the
public interest. Their voices should be united enough
to be heeded when they present alternate plans and list
the consequences of each on public issues. The advice
of engineering professionals will be effective only if
the public has absolute confidence that they will use
their expertise unbiasedly" (62).

Once the constraints are defined, the practitioners should be free to practice in an accountable and
responsible fashion without being harassed by the need

94

for preparing endless reports. All managers and leaders know that the only way to operate an efficient and loyal organization is to delegate authority and responsibility as far down the administrative structure as possible, and to hold those so designated accountable. The more that higher management in industry or government interferes with such subordinates, the more waste both generate, and the more they erode political freedom by generating the need for more constraints.

5C. INCENTIVES AND PATENT POLICIES

The existence of mandatory governmental controls on engineering practice is prima facia evidence that both industry and the profession did not regulate themselves enough voluntarily. Acting jointly they did develop such exemplary technical codes as the ASME High Pressure Vessel Code, the ACI Manual of Concrete Practice, the National Electrical Code, various building codes, etc. Almost none dealt with such topics as waste disposal, occupational safety, or resource depletion. The government had to fill this void. However, as it did so, its agencies generated thousands of pages of regulations, environmental impact criteria, etc., many of which interfaced with the voluntary codes which industries and engineering societies had developed. These constraining regulations could be minimized if all concerned adhered to obligations which enhance the general welfare, and which provide the proper incentives.

Industrial incentives would relate to economic policies which favor the generation of capital for investment in more productive systems. The tax structure would play a dominant role so that profit could be maximized. However, monetary profit should not be the only incentive. In a world of finite resources it makes no sense to design for obsolescence just to maximize profit for the near term. Entropywise (45) it would be more profitable materially to design for permanence, and to provide superior maintenance service. Such a change in societal outlook would require the development of some new economic value system, as well as a reordering of corporate obligations.

Similarly, the incentives engineering practitioners will need to innovate and to lead, as well as to fulfill their role as professionals, will need to expand to include 1) public appreciation and 2) peer respect, as well as 3) financial gain.

95

"Greed for money or recognition is evident in all three. However, the satisfaction an engineer derives from, or the love he has for his work, or the urge he has to serve his fellow man also motivate him to continue his professional development.

The desire for greater wealth or power or status is a deep seated natural instinct. The desire to pursue excellence and to serve one's fellow man can be developed by persuasion and discipline. Peer pressure is a necessary binding agent that insures continued professional development. The pressure generates a subliminal hope for peer approval and public appreciation, and a desire to develop professionally for fear of failure of one's practice through carelessness or ignorance, or because of concern for the technological impact of one's designs on the public.

Financial incentives could be improved if patent laws and unified professional policies guaranteed employed inventors automatic royalties on their intellectual properties" and/or the right to market their patents if the industries they served chose not to do so" (22).

In fairness to such industries, the inventor should remit some of the profit he may make to compensate them for their previous support of the research. "It should be noted, however, that as early as 1979 many companies like IBM, Dupont, and Exxon had adopted a policy offering some compensation for new inventions. A few companies, like Exxon, had returned patent rights to inventors if the companies did not use the invention within five years" (181,188).

"At present most American firms require prospective employees to sign waivers on patents developed during their employment...Current industrial trends appear to favor keeping new developments as trade secrets rather than to patent them because of the possible infringement by a few "unethical" firms, and of the legal expense to fight such claims. Trade secrets, however, force other inventors to "reinvent wheels."

96

Changes in patent policies and laws could be made through joint action of industry "and an engineering unity organization" that would regard patents as intellectual property rather than as a monopoly. The Department of Justice (DOJ) favors the latter and sometimes regards patents as a restraint of trade and in violation of the antitrust law (72). Changes in patent law should define patents as property and should protect the individual's right to own them. The government should assist him legally in defending his claim, rather than entering cases as a friend of the court against the inventor as the DOJ has done (73)" (22).

In Germany, courts protect the inventor's ownership of a patent without charge. Likewise, some foreign corporations specify predetermined awards for patents, and some countries reward inventors for significant contributions. For instance, Russia paid one of its engineers "over $1,000,000 for his patents, and England awarded Sir Frank Whittle, inventor of the jet engine, £750,000 tax free (73). Meanwhile, U.S. constraints and foreign incentives from 1971 to 1978 resulted in a decrease in patents granted annually to Americans from 56,000 to 44,000 while those granted to foreigners increased from 8,500 to 27,000" (74).

Incentives for inventors and industries alike could be augmented by allowing the government to guarantee restricted loans to entrepreneurs for promising developments (74), as is possible in Japan. "Small industries according to Shumacher's book, SMALL IS BEAUTIFUL, are more socially acceptable (75), and they are three times as efficient as large ones in earning profits on invested capital--but they lack capital. Yet they produce most of the significant inventions. Examples are lasers, computers, insulin, Xerox copiers, and Polaroid cameras. Exceptions are TV systems and the transistor (73)" (76).

Incentives related to public appreciation for the engineer's work and peer respect for his technical contributions provide the two other stimuli. These encourage the practitioner to continue his education life-long, and to render service to society as a professional. Peer respect, like public appreciation, must be earned. Most engineering societies present awards to their members for meritorious technical papers or outstanding professional services. Some of the most prestigious awards are conferred by joint action of the several founder societies.

Public appreciation can be improved by public relation's efforts of the engineering societies which educate the public about the engineer's contributions to society, particularly his service as its advocate on matters affecting its health, safety, resources and environment. It would also help if a set of professional policies were developed for engineering advocacy which strengthen the free enterprise system and political freedom, and that replace governmental regulation with professional accountability.

5D. CONSERVATION VS. WASTE VS. PROFIT VS. WEALTH

Engineering was defined (Article 1B) as "the profession in which knowledge...is applied...to utilize economically the materials...of nature for the benefit of mankind." The definition implies that resources will not be wasted operating inefficient systems nor manufacturing worthless trivia. Actually, the definition is ambiguous because the meaning of words like utilize, economically and benefit can be assumed to suit one's whims.

CONSERVATION

Who for instance, is to determine whether that definition embraces the Christian doctrine of man's dominion over all of the earth's materials as well as of all living things, and whether dominion implies ownership or stewardship? Stewardship seems more reasonable since the engineer is expected to maximize efficiency by using as little material and energy as possible. He is dedicated to conservation, not to exploitation or waste. If so, then who owns everything; those now living, or do future generations have a claim? Society never acknowledged future claims until environmentalists objected to the way our air, water and land was being polluted with industrial and public waste, or to deforestation and urban sprawl. These claims were related to preserving the environment for future generations more so than to the waste of nonrenewable materials like metals or fuels like oil. But, what constitutes waste?

98

WASTE

The conversion of crude oil into fertilizers, pesticides, and automotive fuel is not regarded as wasteful. Neither is the diversion of oil to generate power so that farm tractors can be manufactured. Yet, the overall expenditure of this energy on mechanized farms for agricultural products exceeds their caloric value by a factor of 10 since an Iowa farmer uses about 3000 calories to produce an ear of corn containing 300 calories. A simple peasant farmer produces 10 calories for each calorie he expends in labor whereas an Iowa farmer produces 6000 calories per calorie of his labor. Thus, while the Iowa farmer can outproduce the peasant by 600 to 1 in terms of human labor expended, he uses 3000 calories per ear vs. 30 for the peasant and so is only 1% as efficient energy-wise (77). The extra 2970 calories must be taken from the world's oil reserves and will be unavailable ever afterward.

Energy used in this way need not necessarily be thought of as wasted; it is being used to buy time for research and discovery of alternate sources. If man can perfect solar collectors or nuclear fusion so that he realizes a net gain in energy produced over that expended, his civilization may have a chance of continuing. McKetta (78) illustrates the effectiveness of generating systems for non-renewable and renewable sources in Figure 5A. Within a few hundred years at most, the non-renewable sources will be forever unavailable.

Man has extricated himself from apparently hopeless energy balances before. The widespread use of waterwheels in the 1st century A.D. and windmills in the 10th helped, but it was the use of coal in the 15th and oil in the 19th centuries that enabled man to thwart the Malthusian doomsday forcast (Article 2F). Now we seem determined to consume non-renewable fuels far faster than prudence would dictate. For instance, Congress was persuaded by extremists to prescribe pollution controls for all automobiles made after 1975. Controls did benefit people who lived in those few dense urban centers where smog was a health hazard. But, was there a need for catalytic converters in rural areas? Kits could have been designed so that these converters could have been installed and required on all cars operating in smog-prone cities for more than a week or two. This provision would have exempted tourists, but perhaps not truck traffic. Catalytic muf-

99

flers were of no benefit in rural areas for they reduced fuel efficiency 7.5%, decreased safety by increasing fire hazards, and increased car weight by 200
lb. (79).

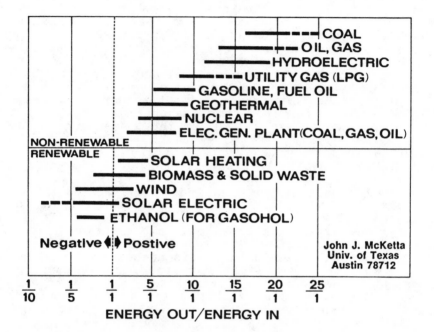

FIGURE 5A - THE EFFECTIVENESS OF ENERGY SYSTEMS (78)

Similarly, Congress was persuaded by common sense
to mandate the development of fuel efficient cars after
the oil shortage was precipitated in 1972. But small
cars are a mixed blessing. They save fuel by reducing
weight, but the weaker structures increase traffic
deaths by 1400 per year. By 1980, compact and subcom-
pact cars accounted for only "38% of all passenger cars
on the road--but 55% of all fatal accidents and 87% of
all accidents involving injuries" (79). The National
Highway Traffic Safety Administration reported that
passengers in subcompacts are from 3.4 to 8.2 times
more prone to fatal injury if they collide with heavier
cars. The decrease in safety is not a benefit. It was
a trade-off.

Any economist could prove that modern farming and smaller cars were profitable as measured in dollars, despite inflation, and beneficial as well. But the resultant process is wasteful in energy units--calories. How, then, is an energy balance achieved to correct so glaring a deficit? The answer, of course, is that the deficit is being supplied by oil. But oil is the legacy we inherited from Mother Nature which she transformed from sunlight and stored over 3 billions of years. It belongs to us and to posterity too.

Were it not for the vast surplusses the American farmer produces, many more people on this over-populated earth would starve. Similarly, smaller cars may reduce operating costs but increase medical costs for the extra accident victims. However, both American farming and automotive transportation waste non-renewable resources. Labor-intensive organic farming and mass-transportation are far more efficient energywise. It takes 8,100 BTU's of energy to transport one passenger one mile by automobile vs. 3800 by mass transit, and 670 BTU's to move one ton of freight one mile by rail vs. 2,800 by truck. The United States already has one mile of road for each square mile of its territory! The American worker's love for his car requires that one of every six jobs be related to the automobile. The car generated enough urban sprawl so that it now takes him longer to drive to work than it did formerly to walk when industry was more decentralized and city populations denser (45). One good thing about smaller cars is that more can be squeezed into traffic jams.

PROFIT

One last comparison of the confusing relationships between conservation, waste and profit should suffice to illustrate the complex societal environment in which the engineer must practice. The discussion here will relate to profit. Modern industry is resource-intensive as well as capital-intensive, and so is highly vulnerable. Yet it is the only hope if the world's poor are to improve their material needs. However, industry cannot supply those needs if it fails to make a profit and goes bankrupt. But, is the free market the only social mechanism that should determine which needs are essential, and which products are trivial? Why should engineering professionals be the only ones dedicated to designing for permanence and conservation rather than for obsolescence and profit? Will Western Civilization's drift from affluence to effluence have

101

to be reversed if the world's masses are to share in this bounty?

This wealth of necessities and luxuries can never be distributed equally, nor is it likely to last long unless man learns to live in greater harmony with his environment. The world is not a closed thermodynamic system since it is continually receiving energy from the sun. That will be free as long as it lasts, but the supply of non-renewable materials like oil, ore, etc. are fixed, and are degraded as they are used; they become unavailable. Open thermodynamic systems like the earth exchange energy in a live energy environment. This theory related to dissipative structures, which won a Nobel Prize in 1977 for Ilya Prigogina, the Belgian physical chemist. Open systems are more complex and generate conditions for greater instability. "The theory ignores the Entropy Law, concentrating on only part of the...process that creates increasing order" (45).

But, in the words of Nicholas Georgescu-Roegen, "Matter, matters." "Matter is continually degrading (also)...recycling only reclaims...a part of whatever matter is used up. We can recycle only matter that is still available." Rubber molecules dissipated from automobile tires are scattered irretrievably over highway pavements (45).

So far the non-renewable resources that industry consumed have been used to buy time and to improve living standards, or wasted to produce needless trivia, as well as to make a profit. If the time is employed to develop a renewable energy source efficiently before some social upheaval annihilates the process, the gamble may pay off. Buckminster Fuller is optimistic enough to predict that it will, since the sun's energy arriving on earth each minute is greater than that used by man annually. He cautions, however, that there can be no hope if every one knew that there was enough for all if we quit fighting over it, but still continued to fight. "What is required are social emergencies of such magnitude...that the failure of existing political leadership will be apparent" to all (80).

WEALTH

Generations yet unborn will think ill of us if we squander their share of nature's inheritance, and will admire us only if we use our intellect to improve their

102

lot and ours. "According to the materialistic view (of the universe), the 2nd Law of Thermodynamics leads inescapably (in a closed system) to the death of everything...This would be true if the universe were material" only (80). Many believe the universe is metaphysical also, and cite the increased understanding we are gaining of the invisible technologies of creation--the atom, the gene and the brain. They believe the existence of intellect inevitably moves everything toward ever-increasing size. They cite the synergistic growth of matter and/or energy particles listed in Figure 5B as proof that hope is possible if man can prevent his own destruction.

System	Synergistic Growth
Inanimate	electron + proton = atom → molecule → crystal → aggregate → planet → star
Animate	molecules + ?* = cell → organism → animal → human → mind → spirit
Human	man + woman = family → tribe → village → city → nation → globepolis

FIGURE 5B - SYNERGISTIC EVOLUTION OF MATTER

Note that each system flips in quantum leaps to a more complex, although possibly more unstable, order (45). If we accept the animate and human sequences as open systems, they will then not follow the increase in entropy associated with a closed system and the 2nd Law to a disordered state. At least they may not until the universe stops the expansion that began with the "big bang" of its origin and reverses its direction into a "black hole." Until then, there will be hope. "(We) are evolving from a phase of separate individuals and nations to an organically interrelated planetary system...Nature is forming a whole system...out of us... just as she did with the cell" (80).

---

*Coacervates, as a watery film, isolated an array of complex organic molecules from their environment to form living cells according to the Oparin hypothesis (81) so that these cells could retain their identity, became stable structurally, and replicate.

The danger we face now may be related as much to ignorance and misuse of power as to resource depletion. "The power structure (political, religious, business) cannot tolerate the success of humanity. Their reason for being derives from scarcity being inevitable" and from fortifying the haves against the have-nots (80). Of course engineering practitioners by themselves cannot, nor should not, endeavor to solve the problems associated with conservation, waste and profit alone. But they should appreciate the complex effect and risks that their practice generates, and try to change those societal constraints which waste resources or endanger safety. The engineer's plight resembles that of a witty Irishman who speculated that, "The path from the cradle to the grave is so beset with perils, 'tis a wonder that any of us live to reach the latter" (78).

As engineers practice and try to satisfy man's wants they must always remember that man's wants are endless and resources scarce. Infinite growth is impossible on a finite planet. The greatest good can be accomplished for the greatest number by choosing how best to utilize scarce resources, and how best to organize society. The principal of contract or voluntary cooperation is to be preferred to coercion. Governmental constraints invariably impede production. The government cannot create wealth, only redistribute it (82).

5E. CONSTRAINTS--CODES, REGULATIONS, INNOVATIONS, UNIONS, AND CUSTOMS[a]

The foregoing discussion implied that the engineer is not free politically, economically or scientifically. His constraints were described by the poet Housman (83) as "these foreign laws of God--and man." The engineer knows that he cannot flout nature's laws without suffering retribution. He seldom does: he understands them. They are simple, enduring and few in number. But manmade laws and the regulations they authorize also affect his practice. The laws are so numerous and confusing that nine learned judges reverse their interpretation of them periodically as the societal environment changes with time.

---

[a]Article 5E is developed here from parts of the material in References 17 and 76.

The confusion arises frequently because the laws which create the regulatory agencies in the executive branch are very hard to formulate in the legislative branch. Legislators, using our democratic process, hold public hearings to develop a concensus, but the testimony presented by special interests is often diverse. Hence, the regulations are hard for public servants to write and to apply, as well as for the judicial branch to adjudicate.

These regulations vary geographically. For instance, automotive emission standards vary world-wide from zero to others some experts deem too stringent. Likewise, arguments persist over whether oil tankers entering certain harbors should be built with double, rather than single hulls, so as to minimize oil spills.

Economic systems also span a wide spectrum (Article 1A). Productive systems can be owned totally by governments (socialism) or by people (private enterprise). Today, all systems are mixed and include elements of welfarism and economic regulation as well. Governments influence the economy and technology by support and regulation. Support includes funding for projects, purchases and research, and expansion of money and credit.

Our over-regulated economic system in 1977 had 68 percent of American industrial executives worried about the ability of their corporate structures to survive the next 25 years (84). They countered the anti-business, pro-union views, taught almost exclusively in schools with paid advertisements. Such ads stated that only free market capitalism is totally democratic; all customers are franchised to vote in the world's markets with their own money.

Technology and economics permeate all political boundaries even though national tariffs and regulations restrain their effects. Technology and economics unite mankind into the one world which Wendell Wilkie endorsed in his unsuccessful 1940 presidential campaign (85). The removal of all economic controls might lead to the increased production and peaceful cooperation the world has been seeking.

CODES

Technology must be controlled by responsible practitioners, or by regulatory agencies, or both whenever

105

the public's safety, health, resources or environment are endangered. There would be little need for these bureaucracies if acceptable professional standards, like the voluntary codes listed in Article 5C were available for all product designs. These develop, however, long after the first prototypes are produced and so are unavailable for innovative creations. Many other voluntary technical standards exist, that are extremely useful to designers, such as those of the American Society for Testing Materials (ASTM) and the American Institute for Steel Construction (AISC). Some mandatory standards--like building codes--also exist which specify procedures rather than engineering principles in great detail. They stifle innovation but were developed because governments refused to restrict design to knowledgeable professionals as they have restricted the practice in medicine and law to those qualified. The application of pertinent scientific principles could suffice in such cases if only knowledgeable professionals had the final authority to make the technical--as distinct from the economic--decisions. Even so, laws--but not regulations--would be needed to enforce contracts which specify adherence to voluntary codes.

The knowledgeable professionals must be held ethically accountable and/or financially responsible if the public is to be protected and to have confidence in the professional's ability (104). Codes, then, can be disarmingly simple, as was the Code of Hammurabi (2124-2081 B.C.). It provided the death penalty for a contractor--or his son--whose building collapsed with loss of life as Figure 5C indicates (105). That regulation accented performance, not code regulations: it was easy to understand; it did not stifle innovation. It pinpointed responsibility.

REGULATIONS

When such responsibility is not permitted and the public's welfare is threatened, mandatory codes and/or regulatory agencies are created. The standards regulatory agencies (Table 1A) develop fill thousands of pages, are frequently irrelevant and always costly to apply. The expense for their direct support, plus the burden they place on all businesses, will be shown later to be five times larger than net profits as based upon sales (17). The regulations were developed to reduce risk and to increase security, but are costly. Senator Jesse Helms remarked in 1978 that security "can

106

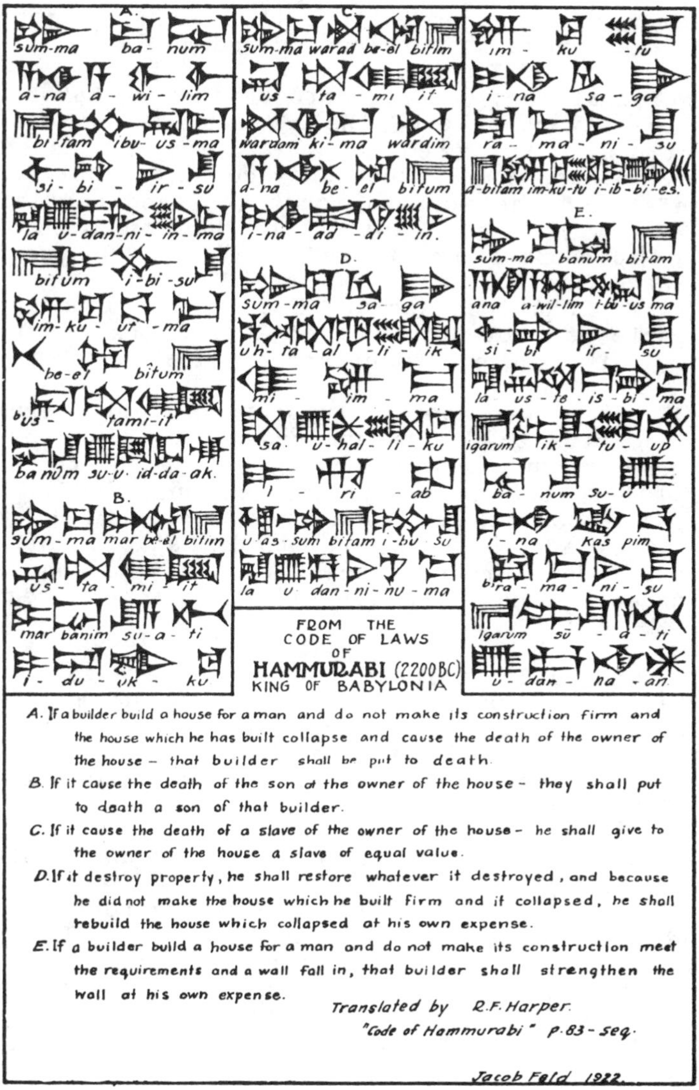
FROM THE CODE OF LAWS OF HAMMURABI (2200BC) KING OF BABYLONIA
A. If a builder build a house for a man and do not make its construction firm and the house which he has built collapse and cause the death of the owner of the house – that builder shall be put to death.
B. If it cause the death of the son of the owner of the house – they shall put to death a son of that builder.
C. If it cause the death of a slave of the owner of the house – he shall give to the owner of the house a slave of equal value.
D. If it destroy property, he shall restore whatever it destroyed, and because he did not make the house which he built firm and it collapsed, he shall rebuild the house which collapsed at his own expense.
E. If a builder build a house for a man and do not make its construction meet the requirements and a wall fall in, that builder shall strengthen the wall at his own expense.
Translated by R.F. Harper
"Code of Hammurabi" p.83-seq.
Jacob Feld 1922.

FIGURE 5C - THE CODE OF HAMMURABI - 2200 B.C. (105)

From CONSTRUCTION FAILURE by Jacob Feld
Reprinted by permission of John Wiley & Sons, Inc.

sometimes be provided at too high a price, and that what people need is relief from protection." He said, "We now have more regulations than we want, more than we need, and more than we can afford."

For instance, General Motors Research Laboratories sometimes divert half of their resources to meet these regulations. Dow Chemical, in 1976, estimated that one-half of the compliance money could be justified for the safety of the workers, customers and public. The rest was "beyond good scientific manufacturing, business or personnel practices" (86). Steel companies would have had to spend $1.2 million per steel worker to satisfy OSHA inspectors. They accomplished comparable employee protection for $42 per worker using ear plugs, monitoring equipment and physical checkups (86).

Direct governmental support of all regulatory agencies, plus the indirect burden they place on society, have been conservatively estimated to average about 10% of the gross national product (GNP). Similarly, the Council on Wage and Price Stability concluded they add 3/4% annually to inflation. About 95% of these Federal regulations were never passed as laws by Congress. They were devised by what some regard as invisible dictators.

If we could eliminate the burden this governmental interference places upon individuals, and assume that perhaps half of the regulations imposed on industry are beneficial, we could return the equivalent of about 5% of the GNP to industry as a tax incentive or tax-reduction (Table 5A).

That 5% rebate would represent enough capital to allow industry to modernize its plants and to reward its employees with incentives for greater innovation and adequate royalties on their patents. If one recalls that 65% of the GNP represents spending by the private sector, and that the average net profit for business is 3% of sales, the estimates show that a 5% rebate would almost triple industry's net profits (Table 5B).

Industry needs that capital badly for modernization and expansion, especially since robots are now being introduced worldwide. Machine tools in the United States have an average age of 35 years compared to only 7 years in Japan. Our steel industry is using some furnaces dating back to 1890, whereas the competitive

108

TABLE 5 - ESTIMATED COST OF GOVERNMENTAL REGULATION
vs. INDUSTRIAL NET PROFIT (17)

TABLE 5A

**Estimated Direct and Indirect Costs
of Governmental Bureaucracies**

General Accounting Office	5% GNP
Ford Administration	10% GNP
Conservative economists	20% GNP
Assumed average	10% GNP
Assume ½ is essential[1]	5% GNP

[1]Return balance to industry.

TABLE 5B

**Industrial Net Profit vs.
Governmental Regulation
in Terms of the GNP[1]**

Spending—private sector	65% GNP
Spending—public sector	35% GNP
Assumed cost of bureaucracies	10% GNP[2]
Net industrial profit—estimated at 3% of sales or $0.03 \times 0.65 = 0.0195$ or about	2% GNP

[1]Exclusive of transfer payments like Social Security, unemployment insurance, government pensions, etc.

[2]If only half of this cost were eliminated *and* returned to industry, its NET profits would almost triple!

109

Japanese have already scrapped some "old" steel plants built in 1955! Imposing import tarrifs on foreign steel will not reduce the need for American industry to modernize.

Similarly, higher taxes, compliance costs and lower profits have forced American industry to export jobs. American overseas plants now account for 75% of production for the electrical industry, 33% for our chemical and pharmaceutical industry, and 25% for our automotive industry.

INNOVATION

Industry must have enough venture capital to market new and improved products, and enough profit to reward its inventors if innovation is to flourish. The risk of marketing a new product in a new market has been estimated at 20 to 1, whereas for an old product in an old market the odds for profitable operation are reduced to 1 to 1. Industries, whose single goal is profit, will opt for the latter. That objective retards progress and can ultimately destroy at least a segment of an industry. That is what happened to Baldwin-Lima-Hamilton when it continued to manufacture steam locomotives and to ignore the Diesel engine.

Innovation embodies the right to fail, or to reap high rewards. From 1945-1974, new industries like IBM, 3M and Xerox had annual compounded growth rates of about 16% in sales and 11% in jobs. Mature steel and chemical industries averaged only 8% and 2% respectively. Bell Telephone's transistor, invented in 1948, was really marketed under license by some Bell scientists who left its employment. The character of radios, calculators and watches changed dramatically, and so did their manufacturers. Raytheon, GE, Friden and a host of others were jolted.

Curiously, sometimes it has been our bureaucratic agencies that have forced innovation on a reluctant industry. The change from drum to disk brakes on American autos is an example. Our huge investment in automated machines making drums might have overridden adoption of the successful European disc, had it not been for new Transportation Department standards.

110

No one needs to remind engineers that regulatory constraints reduce productivity by diverting resources and labor. Engineers are not as aware, however, that productivity could also be increased by imposing some governmental restraint on labor unions. One can only wonder why monopolies and contract violations are prohibited by government for business but not for labor unions. Labor unions sometimes restrain production particularly in the building trades where they prevent any interchange of incidental job functions between skilled laborers. Unions developed a century ago because industry then was exploiting adult and child labor. Unions have benefited labor and have succeeded only because they can halt production by striking. But strikes are of little use to engineering staffs because production can continue for a while even if engineers stop working. Furthermore, many engineers aspire to management roles, and these are excluded legally from collective bargaining. Only in some college faculties have unions been successful in unionizing engineering professors. There they are submerged in the whole faculty and prevented from constituting a separate bargaining unit for professionals, as is the case for medicine and law school faculties or professionals in industry. Figure 5D shows the growth of collegiate unionization activity from its inception in 1971 to 1981. About 25% of college faculty were unionized in 1980. A U.S. Supreme Court ruling on the Yeshiva University case in February 1980 is thwarting this growth. It ruled that faculty exercised management functions and were ineligible for collective bargaining. The management functions included supervision over curricula, admission, tenure and faculty selection. By September 1980, 37 unionized colleges had refused to bargain successor agreements.

Collective bargaining laws, limiting the power of labor, received considerable impetus in 1947 when the Taft-Hartley Act was passed over a presidential veto. This law specifically allowed professionals in industry to bargain separately. They had the choice of forming their own collective bargaining unit, of joining the larger one applying to all workers at their plant, or to remain free.

Somehow, unions were never able to convince many engineers employed in industry to join the movement. About 5.4% of these 830,000 practicing engineers were

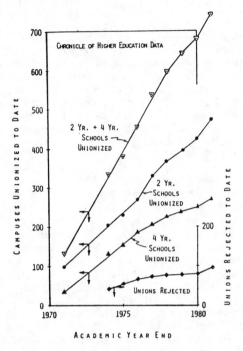

FIGURE 5D - CAMPUS UNIONIZATION ACTIVITY

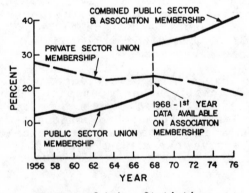

Bureau of Labor Statistics

FIGURE 5E - PERCENT OF WORKERS UNIONIZED

represented by collective bargaining units in 1976. "More significantly, only 2.4%, or about 20,000 are actually union members" (87). Their reluctance may stem from an unselfish dedication to serve mankind more effectively, as free of constraints as possible. Their reluctance may also stem from a more selfish aspiration to ultimately join the ranks of management which are immune from unionization. Organizing engineers into labor unions, or into a unified professional society, is a seemingly impossible task.

Strangely enough, the attitudes of managers toward collective bargaining are changing. According to an American Management Association survey in 1971 of 6000 executives, 18% of the 1100 who responded were willing to join a union. Another 17% said they would consider union membership. Half said they favored a change in labor laws to permit unionization of managers in supervisory positions and 75% favored informal sounding board organizations (88,89).

The objectives of unions and of professions involve loyalties in an inverted order as Table 5C shows. Society has generally rewarded the professional for his unselfish service with better than average creature comforts as well as respect. Action based on individualism, not collectivism, has characterized his service. Society does not willingly confer such rewards on those who also serve it when their primary loyalty is to themselves. Unfortunately, the great mass of workers have had to organize into collective groups to realize even a fair share of the wealth they created or personal service they rendered. Had labor's service been as unselfish as that of the professions', its members probably would not have fared as well. There is too much greed in the world. The workers had no choice but to unionize. In so doing, they were forced to consider their own welfare first. Collective bargaining, and ultimately the strike, were their only weapons. All too frequently, strikes penalized the public as much as the strikers.

Collective action had another unexpected and unwanted effect. It made all members of society insensitive to each other. Seldom do they act as individuals now and come to the aid of the injured, the aged and the poor. Most of us have "passed the buck" to amorphous health or welfare agencies rather than to get involved. But collective action on the part of labor has numbed society's feelings no more than has individual

113

specialization of the professionals, as sociologists
Rudoff and Lucken (28) pointed out.

TABLE 5C--LOYALTIES OF MEMBERS

OF PROFESSIONS		OF LABOR UNIONS
1st	Public	1st	Self
2nd	Employer-Client Peers	2nd	Union-
3rd	Peers-Profession	3rd	Employer
4th	Self	4th	Public--?

Professional independence, like academic freedom,
implies the existence of unfettered individual effort.
If engineering ever congeals into a mature profession,
it will require the collective action of individuals
dedicated primarily to serving society with their technical expertise and leadership. No mature profession
will use collective bargaining to enhance the personal
wants of its members, be they graduate practitioners or
faculty. Surrendering of a faculty's individualism to
collective bargaining will make its members more insensitive to their fellow men, and more concerned with
their own needs than with those of society. That attitude is bound to be transmitted to the students. It
does not embrace the concepts of professional service.
Contractual service by faculty would eliminate the possibility of their unionization (89).

Although unions, in 1976, attracted 19 percent of
industrial workers and 39 percent of public employees,
as Figure 5E shows, they managed to attract only 2.4
percent of employed engineers for the reasons cited
above. These engineers might never have joined unions
had their professional societies been able to do more
than they did to enhance employment conditions, economic rewards, and peer support through professional ombudsmen, especially when the engineers were endeavoring
to protect public safety and to uphold ethical codes.

CUSTOMS

The foregoing discussion of the societal environment in which engineers must work should leave little
doubt that the host of conflicting customs make the
world seem to be hopelessly complex. However, the same
can be said about the natural environment. Yet the engineer has learned to adapt that to the comfort and
convenience of his fellow man. He has accomplished

114

feat by applying the few scientific principles he needs. His success can be measured by the economic and social incentives provided by society for his service.

Actually, when one considers the volume of production, American industry has had a remarkable record for manufacturing safe products and for rendering safe services when measured in terms of the gross national product (GNP) and the total population. Yet, the record is far from perfect if malpractice cases, liability suits, consumer product recalls, collapsed astrodomes and ill-designed drainage projects and municipal incinerators are considered. The public pays even though some of the losses are covered by insurance, for insurance merely spreads the loss among all consumers. The public pays to maintain the extra courts needed to handle the additional legal cases that burden our judicial system.

In 1970, for instance, the President's Commission on Product Safety stated that "The exposure of consumers to unreasonable consumer product hazards is excessive by any standard of measurement." It reported that some 20 million Americans are injured each year as a result of incidents connected with consumer products, and an additional 2 million are disabled on the job (93,94). This does not mean that engineers in industry are poorly qualified or careless or irresponsible. Many of these incidents are the fault of individuals injured or killed rather than the products. It is not always possible to anticipate how the consumer will misuse a product. Nevertheless, it behooves designers to make products and services as fail safe--and fool safe--as possible to eliminate costly law suits. In 1969 over 300,000 such cases were filed, a three-fold increase over 1968 (95).

The main point to emphasize here is that although all losses are small compared to our total GNP and population, the cost for regulation, insurance, legal fees, injury, property damage and lost tax revenue may exceed industries' net profit (based upon sales) several fold!

The future will demand more of the engineer if technological progress is to continue. He can continue to improve his service if incentives can be expanded, if controls can be removed, and if his professional objectives can be awakened and broadened. But how can present customs be changed? First, we can start by ex-

115

changing costly governmental interference for increased industrial profit by transferring final authority and professional accountability for technical decisions to individual engineers in responsible charge. Second, we can recapture our innovative advantage by using tax relief to finance plant modernization and patent royalties for inventors. Third, we could weld the governmental-industrial-labor spectrum into a social force whose esprit-de-corps emphasizes enhancing the public good while providing each contributor with optimum rewards. All three would improve our worldwide competitive advantage and help to restore our balance of payments.

Although the suggested changes in customs embody legalistic, moral, social, and technological elements that would be difficult to implement, it need not be inferred that, as Solzhenitsyn's Harvard University commencement address implied, "the West has lost its courage." Its courage may be only dormant. Our apathy encourages defense of human rights more than respect for human obligations. If as engineers, our "paramount responsibility is to protect the public health, safety, and welfare," we must assume individual accountability for our engineering practice, and not allow technical requirements to be usurped by governmental regulation nor compromised by managerial decisions. Regulations are no substitute for ethical behavior.

6

Engineering Education[a]

6A. THE OBJECTIVES - TECHNICAL LITERACY,
COMPETENCE AND LEADERSHIP

Those who study the history of education will soon
notice two divergent objectives of the system; i.e., 1)
the urge to discover and the need to transmit the practical knowledge required by people to survive and to
progress in primitive or industrial eras, and 2) the
desire to contemplate life's meaning and to appreciate
its beauty. Both are needed for today's society, especially if we hope to develop leadership and competence
in educated and cultured engineering professionals.
Their ability to comprehend research publications scientists produce so as to be able to apply relevant discoveries is essential for competent societal service as
well as for their own advancement.

Professionals cannot master their expertise today
without a prolonged period of formal study. New knowledge increases faster than older knowledge is proven
obsolete or discarded. Even though knowledge can be
distilled and presented more effectively by great
teachers, periods of formal professional study have
been lengthened throughout history. But so has man's
life expectancy. There appears to be a limit of about
one-third of one's lifetime that can be devoted to formal study for professionals, although continuous life-
long study thereafter is a necessity to upgrade and to
update one's competence.

Criteria for professional education cannot be
formulated unless three specific definitions are accepted: 1) that of a profession, 2) that of the education

---

[a]Extracted from the Report of the NSPE Workshop on
Professional Schools of Engineering March 1978 (64).

to be mastered by those who are designated as professionals, and 3) that of the optimal educational environment required. These criteria often provoke debate, sometimes with great fervor.

FIRST, let the accepted definitions for professions and the statements of their purpose be those stated in Articles 1A and 1B. These are summarized here for convenience. Dean Pound condensed them to "the pursuit of a learned art in the spirit of a public service." This objective has been incorporated into engineering codes of ethics so that engineers must "hold paramount the health, safety and welfare of the public in the performance of their professional duties," and must report to the proper authorities any violations of that canon of the codes by any person or organization.

Let it also be remembered that professional status is conferred by the public upon those groups of practitioners whose knowledge requires a long period of formal education, whose practice provides beneficial public service, and whose expertise is unknown to laymen but has a mystique often feared by the public. Professional stature cannot be assumed by any group, nor can any faculty assume arbitrarily that they teach in a professional school.

SECOND, let the quality and quantity of formal education to be mastered be determined by both educators and practitioners so that the proficiency of graduates can meet criteria for professional practice in either practice-oriented or research-oriented or management-oriented careers.

Graduates of professional schools 1) must be competent to practice at an adequate technical level, 2) should be able to read all technical literature in their specialty, 3) should be broadly educated in all engineering sciences, management, and economics so as to be able to provide leadership in the industrial, governmental and societal sectors, and 4) should have assimilated sufficient culture so as to feel at ease among other educated peers or scholars. The importance of training in management and leadership cannot be overemphasized. Either engineers will be managers, or they will be managed, as they work on technical missions: they can either lead or be led (90). In the future, engineers will have to provide collaborative leadership in society to formulate technical missions,

118

and to control future crises, if our free enterprise system and political freedoms are to survive.

THIRD, it is doubtful whether those objectives can be fulfilled beyond the 1970's with less than six years of formal college study. Further, the esprit de corps needed by professionals, when they are confronted by selfish groups endangering the public's health and safety, can hardly be imbued in composite student bodies of large universities. Such dedication can, and has been, transferred throughout history in smaller professional schools for soldiers, priests, physicians and lawyers. Professional competence for practice-oriented careers could likewise be developed for engineers in distinct professional schools whose primary objectives are to graduate professionals willing to assume a stance as the public's advocate and lobbyist's adversary on matters affecting the public's resources, environment, health and safety. Such schools would have to indoctrinate their students in these professional ideals, and in the need to abide by and to enforce a code of ethics. Few engineering schools anywhere in the world now measure up to all of these criteria. The United States could initiate that trend for many more. The coming post-industrial world will need such leaders desperately.

If the foregoing definitions and general criteria are accepted, one can then note the perturbations in the educational efforts throughout history as the emphasis shifted alternately to and from a study of the practical knowledge needed to survive and to create wealth, and to and from a study of culture and of contemplation. The search for new knowledge can be applied to either. The following history illustrates this unsteady state of human endeavor as well as the foundation of modern engineering education.

6B. EARLY EDUCATIONAL HISTORY

Ever since the dawn of history man has found it imperative to learn in order to survive. He enhanced his security by teaching his offspring two things--the need for authority and group stability, and the need for skills and individual specialties. Families were the first educational units where children were taught fundamentals like speech, crafts, fishing, loyalty and discipline. A ceremony at the end of the child's puberty signified the successful completion of this primitive education.

The home was the primary educational center for ancient Jewish families, where education has always been emphasized, and where the family has been held responsible for its fulfillment. Even so, in Palestine, elementary schools for children were not required until 64 A.D. Among Hindus, the family effort was supplemented by teachers and apprenticeships.

Later, as family units clustered into tribes, and tribes into city-states, occupations diversified into social strata to include soldiers, priests, nobles, workers and slaves. As trades developed among the workers, knowledge was transferred from father to son. There was little chance for upward social mobility except after the Renaissance in Europe, or in ancient China which selected talented youngsters for higher education as early as 200 B.C., and officials for public office by examination at least by 500 A.D.

Originally, knowledge was transferred among commoners and slaves by memorization and by imitating teachers. There were no textbooks. Imitation was supplemented by apprenticeship in special trades, and for individuals selected for higher occupations.

All of this early primitive education was vocational in nature, emphasizing "how," not "why." Gradually, as city-states developed, and irrigation coupled with agriculture began to provide a surplus of food, of time for the development of specialties, and of leisure for an elite, vocations began to assume the status of professions for the ministry, the military, law and medicine. Human society was revolutionized by establishing governments needed for the common defense, and to maintain order and to distinguish custom from law. The several specialty groups we now categorize as learned professions were born. But there were seldom schools in primitive times as we know them today.

Early Egyptian education of 3000 B.C. for their professional elite included writing, mathematics, astronomy, music and science. Several Greek scholars studied there about Plato's time. Early Greek education was restricted to two essentials: professional teachers and students. Classes were held in groves or temples. There were no school administrators then.

Plato, like his master Socrates, was among the first to give a regular course extending over three or four years in a fixed place--an academy at which no fees were charged, no examinations were given, no degrees were awarded and no licenses were required for the teachers. Education was compulsory in Sparta and available only for those who were born free--for an elite; not for slaves. Aristotle, for instance, tutored Alexander the Great in 342 B.C.

Professional teachers were of three types: those who taught reading, writing and arithmetic; those who taught music to boys 7-15; and those who taught physical training until the boys were 18, when they entered the military for two years. Boys were taken to these academies by slaves. Girls were not educated in these essentials, nor in the arts.

Athens became a cultural center as it grew in political power. It attracted sophists whose teaching developed two schools: a philosophical school, where later Socrates, Plato and Aristotle taught mathematics, science and dialectics for a life of contemplation; and a rhetorical school, made famous by Isocrates, which emphasized grammar, speech and literature so as to prepare students for (professional) public service. By the 4th Century B.C., these two Athenian schools developed into a university attracting students from distant lands. Later, the university at Alexandria took the lead, but its library of 700,000 volumes was burned in 47 B.C. Scholars of these eras recognized how important education was to the stability of the State, and selected students according to ability.

Early Roman education was described in Article 4E as being more practical. It included the basic study of the three R's and emphasized leadership; i.e., professional education like military training for some, and oratory and management of public and private affairs for others. Both programs were combined later but their importance deteriorated after the Roman Empire declined and barbarians overran Europe in the 5th Century A.D. Christianity began to flourish. Monasteries preserved learning in a status quo state as scholarship declined.

However, the educational dichotomy created in the early Greek academies, emphasizing elementary fundamentals and fine arts but ignoring vocational studies, set a pattern that persisted until the Renaissance. Hippocrates' teaching of medicine in the 4th century at the Temple of Cos may have been an exception. Plato considered thinking man's noblest effort. His contemporary, Xenophone (434?-355? B.C.) regarded the practice of mechanical arts to carry a stigma; work was for slaves. That practice is still so regarded in much of the modern world, and has been one deterrent to public acceptance of engineering as a profession.

6C. EMERGING MEDIEVAL EDUCATION

Many clergy of the emerging Christian religion benefitted from the Jewish-Greek-Roman learning. They soon realized the need to educate their professional clergy and developed a fixed curriculum based upon the "severn liberal arts" of grammar, rhetoric, dialectics, arithmetic, geometry, music and astronomy. These curricula seemed to favor seclusion in Monastic life rather than service as Parish priests. Monasteries emerged about the 4th century in Asia Minor. Much later, in 1594, some, like the Jesuits, were founded as service oriented, teaching orders. Meanwhile, the early Christian curricula lacked the former Greek innovative inquiry and even purged Pagan literature in 529 A.D. A serious decline was finally halted by Charlemagne in the 7th century by ordering the clergy to found schools in cathedrals and monasteries.

Besides raising the educational level of most of the elite and some of the masses, the Medieval church also effected a profound change in society from the 4th to the 14th centuries. Without condemning slavery, it nevertheless created a religious desire to distribute "nature's bounty" among all parishioners. Slavery did disappear and the number of small landowners expanded immensely. The number of unpropertied "proletariat" was reduced substantially, and did not increase again until after the Industrial Revolution. But above all, the seeds of individual freedom had been sown.

The dual scholastic and social effort led to two developments: the emergence of universities in the 11th century in the form of professional schools, and ultimately to the industrial and political revolutions of the 18th century. Salerno became a recognized medical center in 1050 A.D., Bologna was chartered in 1158 A.D. to teach law, and the University of Paris was opened in 1175 A.D. with programs in cultured arts and in professional theology. Programs were divided into apprentice, masters or doctors levels. The titles of magistrate, doctor or professor were considered equal.

These institutions of higher learning were formed by student guilds--or studiums--as at Bologna, or by master guilds, as at Paris. The studium generale ultimately became complete universities. Graduates became teachers, lawyers, doctors or ministers; i.e., professionals. Oxford (1168) and Cambridge (1209) were studium generale.

Modern universities also grew out of the followers of great teachers as had been the case for the Greek academies. For instance, Peter Abelard (1079-1142), a historian and philosopher, was educated in a cathedral school but flouted their teachings. He attracted throngs of students in Paris, but was imprisoned twice as a heretic. Similarly, Irnerius of Bologna, was famous as a teacher and revived an interest in Roman and Canon law.

The feudal nobility, however, were tutored mainly in palace schools in knightly arts, music, sports and manners. Guilds were organized to provide vocational apprentice training for trades. Thus, professional and apprentice schooling was well established by 1200 A.D., but elementary schools were unavailable for the masses until the late Middle Ages (about 1500 A.D.).

The broad philosophical programs common today in universities had become dormant after the Greek era, and were revived only as the Renaissance emerged about 1300 A.D., particularly after the Reformation about 1550 A.D.

A great revival of art and learning erupted in the 14th century, and led to the establishment of scientific academies by the 16th century. The research in Greek and Roman literature, and emphasis on experimentation was resisted by universities, since their curricula were oriented to a study of theology, or to specialties in medicine and law. However, at the elementary school level, Luther strongly supported education so that all Christians could learn to read the Bible. Research and innovation began to triumph.

The foregoing paragraphs illustrate the different educational patterns that evolved for trades and guilds with their apprenticeships, and for the learned professions of law, medicine and the ministry in separate schools. None of these efforts were connected with the early Athenian academies which specialized in intellectual studies like philosophy and shunned vocational education. Their liberal studies were meant for leisure and contemplation. Early Roman education, it will be remembered, was more practical. It emphasized citizenship and military training for some and oratory and the management of public affairs for others.

6D. MODERN PROFESSIONAL SCHOOL DEVELOPMENT

The first modern universities emphasizing secular and liberal studies were founded at Halle (1694), Gottingen (1737) and Erlangen (1743). The Reformation movement of central Europe broke the previous traditional pattern for programs of only theology, law and medicine, and allowed the formation of liberal, secular ones of science, philosophy and the arts.

In America, Harvard (1639) and William and Mary (1693) were founded principally to educate clergy, but law schools were chartered there in 1817 and 1779 respectively. Yale (1701) also began to "fit youth for public employment in church and state." Although these career programs were organized as professional curricula, supporting courses in history, government, science, mathematics and literature were soon included so as to produce leaders who were cultured as well as educated for their professional practice.

Many separate law schools also flourished, the first being that of Tapping Reeves in Connecticut in 1784. Separate medical schools were founded in 1754 at what is now Columbia University, and in 1765 at what later became the University of Pennsylvania.

Formal education in engineering began in 1716 when the French engineer, Jean Rodolphe Perronet, was authorized to instruct "designers in sciences and practice needful to...bridges and highways" for the Corps des Ponts et Chaussées. This effort led to the first school of engineering in 1757--the Ecole National des Ponts et Chaussées. Later, others were created like the Ecole des Mines (1778), Ecole des Arts et Metiers (1788), Ecole Polytechnic (1794), and Ecole des Arts et Manufactures (1829). Technische Hochschule followed soon after in Germany, the first at Karlsruhe in 1825.

Russian engineering education started about the same time as the French. Peter the Great (1672-1725) ordered the teaching of mathematics, navigation, artillery and utilitarian purposes. This led ultimately to the establishment in Russia of the Academy of Sciences

(1725), of the Military School at Gentry (1731), and of the Institute of Mines (1774).

In England, nonconformists, who were excluded from universities, established clandestine dissenting academies to teach modern science, technology, economics and management. Faculties at Oxford and Cambridge resisted the introduction of subjects like natural history, geology, etc., and were ultimately challenged by universities in London (1827), Manchester and Birmingham. However, apprenticeship remained the favored educational pattern in engineering there until after 1850 although several mechanics institutes were founded.

In the United States, the Military Academy at West Point, New York was founded in 1802, but formal professional curricula in military engineering really began there in 1817 when Sylvanus Thayer patterned them after the French schools, which he had visited. Somewhat similar professional military academies were formed at St. Cyr in France, and at Sandhurst in England. Separate civilian schools of engineering were started at the American Literary, Scientific and Military Academy (now Norwich University) in 1813 by Alden Partridge, at the Rensselaer School in 1824 by Stephen Van Rensselaer, and at the Gardiner Lyceum in Maine in 1822 by Robert Gardiner. The last one had programs in agriculture, arts and engineering, and was the forerunner of schools established after the Morrill Land Grant Act was passed in 1862.

Separate engineering curricula were added at Harvard and at Yale in 1847, and at Dartmouth in 1851. Thereafter, most American engineering schools were added as separate curricula at existing universities. By 1862, there were about 12. The Morrill Land Grant Act established many more to teach agriculture and the mechanic arts, so that by 1870 there were 70 as Grayson noted in his excellent summary (91). The growth thereafter was phenomenal. By 1981 there were 250 schools with ECPD (now ABET) accredited curricula. Few, if any, of these undergraduate curricula have as their principal objective the building of competence in and preparation for practice at a registered professional level, as is usual in law and medicine.

Graduate study in engineering was quite meagre until about 1930. It began to increase after World War II, and took a dramatic jump after the Russian satellite Sputnik was launched in 1957. However, almost all of this educational effort was research-oriented, not graduate practice-oriented, even though only 15 percent of job opportunities involve research assignments. Whether any of these graduate curricula can be considered as professional depends entirely upon one's definition.

6E. EVOLVING ENGINEERING EDUCATION

The foregoing paragraphs described the struggle schools dedicated to professional objectives had as they developed. Often engineering education was not regarded as a scholarly or even respectable pursuit. It was believed to lack culture. It suffered also by comparison with other professional programs which required previous college study as a prerequisite.

In the defense of engineering education one might add that no other baccalaureate program is as broad culturally as is engineering. Specialists in, say, literature alone are not cultured if they know nothing of other disciplines. Nevertheless, the social-humanistic and economic stems in engineering education for professional engineers could be improved immensely if these courses were integrated with technological history. Instead, these stems are usually taught as entities, designed only to produce scholars who then reproduce other scholars in their own image. Some of these disciplines have already grown beyond the needs of the market and society if one considers their purpose is only to educate teachers, and not to transmit knowledge.

Engineering educators are still searching for the type and minimum amount of social-humanistic studies needed for an engineer's "liberal" education. Obviously, the engineer will need to study selected humanities to appreciate the heritage of the ages--the best in music, history, literature and art, and to enjoy his leisure. But he will also need to know something of political science, sociology and economics to practice his art, and to contribute his expertise to the control of public crises, like those involving energy, pollution, inflation and resources, if his societal leadership is to become effective.

The search for optimum curricular content for any professional program is a never ending one, especially in science related fields. The ever present danger is that it may become too irrelevant as happened to reli-

126

gious education prior to the Reformation and to American engineering education prior to World War II.

Practically the whole emphasis of American engineering education has always been on the technical content, with very little accent, if any, on professional obligations to the public. Social-humanistic subjects have always been included, but their total seldom exceeded 20 percent of the baccalaureate effort. It is likely that if most early American engineering schools had not been founded as separate undergraduate curricula in established universities, but as graduate schools, as was later the case in medicine and law, the engineering profession would have served society even better than it has. In 1867 Dartmouth did develop such a program, capping three years of general studies with two years in a professional school.

American engineering education was developed by educators rather than by practitioners. The formation of most engineering technical societies lagged behind the introduction of such curricula in schools. This unique educational effort was based upon technical courses borrowed from France, shop courses from Russia, fine arts from England and graduate study from Germany. However, it was the addition in the United States of laboratory work, of management courses, and of cooperative experience at some schools that gave it added technical strength. Nevertheless, the consistent lack of emphasis on professionalism, ethics and leadership, even through doctoral study, has prevented engineers from serving society with greater efficiency. Had they been more independent of and less subservient to demands of the public and management, it is likely that the technical crises society now faces would have been less severe. They would have been more conscious of the public's health, safety and welfare if they had assumed full accountability if not full responsibility for their practice, and been less inclined to design for obsolescence to waste material and energy.

Although educators may have been remiss in not including instruction in professionalism and ethics, and in not providing sufficient breadth and depth to curricula before they designated graduates as engineers, they did at least provide periodic curricula accreditation at 6 year intervals after 1934. Educators did not accredit curricula in perpetuity.

Practitioners, however, were remiss in licensing graduates for life after a short internship following graduation, or even without any collegiate education at all. Recent public disenchantment with all professions has forced them to consider periodic re-examination for licensure.

Thus, both educators and practitioners have been remiss. Not all engineering curricula available to the public have been accredited, as is the case for law and medicine. Neither do all engineers who practice need to be licensed. Those sheltered in industry are exempt; consultants are not. Undoubtedly these exemptions have resulted in the development of faulty products and systems by a few industries more interested in profits than public health and safety. If educators and practitioners alike would eliminate the deficiencies cited above, many more engineering professionals of genuine stature would appear. First, however, schools will have to modify their objectives to produce such graduates, or present practitioners will have to be retreaded in that image. Their need is critical now in society. That need will increase steadily as, and if, this technological civilization continues.

Engineers of the future who aspire to practice as professionals, and to serve as leaders who establish societal technical missions, will have to assume the role as the public's advocate and the lobbyist's adversary. Engineers must be educated for that role or they, and all other professionals, will lose their professions and the chance to serve society as free men.

6F. ENGINEERING PROGRAMS AND CURRICULA

PROGRAMS

Periods of formal study for all learned professions, except engineering, have increased over the past century to an average of six to eight years of post-secondary-schooling when the degrees conferred designate the graduates as professionals. Figure 6A (7,96) illustrates how these programs increased since 1900 for seven professions. For instance medical education increased from 3 years of post-high school, *professional study* in 1900 to 8 years by 1972, 4 years of which are *pre-professional* baccalaureate programs in the arts, sciences or engineering. Similarly, law increased from 2 years in 1900 to 7 years by 1972, 4 years of which are at the B.S. or B.A. level. All of the other disciplines shown in Figure 6A except engineering experienced similar expansions.

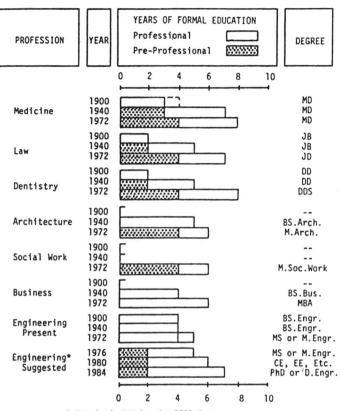

Figure 6A - Structure and Length of Professional Programs (7)

The engineering educational requirement of 1900 was the most rigorous! It was the only one then to specify a 4 year B.S. degree as the *first designated degree* for entry into the profession. Yet in 1980, it was the only profession still doing so despite the exponential growth in knowledge, especially since 1940 (see Figure 2A).

Several studies were made to remedy what some engineering societies considered was a deficiency. Among those sponsored by what is now the American Society of

Engineering Education (ASEE) were the Wickenden Report (97) of 1924, the Hammond Reports (98) of 1940 and 1944, the Grinter Report (99) of 1955, and the Goals Study (100) of 1968. The first two were concerned mostly with broadening the cultural content and recommended that it constitute at least 20% of the program so as to make the student keenly aware of the heritage he acquired from the past as well as of the inheritance expected from him in the future. These early Reports, however, also emphasized the need to acquire a sound base of science and mathematics so as to apply them in practical ways. Similarly, Everitt (102), in 1944, suggested the need to emphasize ethics and synthesis as well as analysis.

The Grinter Study (1951-55) re-emphasized the cultural criterion but was more concerned with the need to upgrade the scientific and mathematical foundation, to strengthen the design requirement which distinguishes engineering from most college programs, and to recognize the obligations of the profession to society. These three became evident in World War II when western democracies had to rely to a greater extent upon scientists to develop such technological advances as radar and nuclear fusion to win that struggle more quickly.

The Grinter Report (1955) provided the criteria for the long overdue introduction of these changes. At first, these were resisted, but the launching of the Russian satellite Sputnik in 1957, and their ensuing adoption as ECPD accreditation requirements accelerated their acceptance. New courses were introduced and obsolete ones were eliminated. Unfortunately, the essential credit hours required for the engineering baccalaureate were reduced 10% in the early 1960's during a slump in engineering enrollment (see Figure 6B) (101) so that these programs could compete for students with those in arts and sciences.

The Grinter Report did not recommend any reduction or extension in required credit hours or program length, but concentrated instead upon quality and purpose. New faculty appointments beginning in the early 1960's had to be restricted to engineering or science doctorates in order to upgrade the program content. Unfortunately, too few had prior practical experience as well (27). However, they did recognize the need to expand engineering graduate study. Thus, it was only natural that both undergraduate and graduate programs developed with much more of a research-oriented empha-

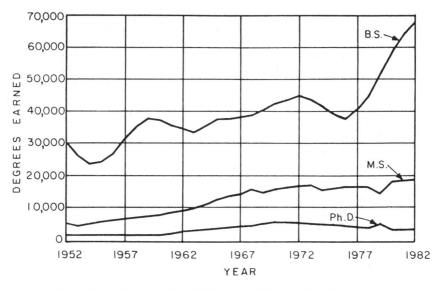

· AAES ENGINEERING MANPOWER BULLETIN, DEC.1982
(101)
FIGURE 6B - ENGINEERING DEGREES EARNED

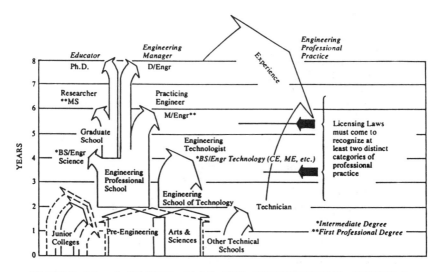

FIGURE 6C - A MASTER PLAN FOR PROFESSIONAL SCHOOLS

AT TEXAS COLLEGES (96, 103)

131

sis than the earlier practice-oriented ones. Both, of
course, are needed; neither should exclude the other,
particularly since at least 15% of employed engineers
engage in research. Both should also embrace some
study in engineering management and societal leadership
so that graduates can fulfill their profession's public
purpose adequately. Instruction in management and
leadership, even through doctoral programs was, however, almost entirely lacking even though engineering
practice accounts for one-third of engineering management functions as the 1968 and 1981 surveys shown in
Table 6A also indicate.

TABLE 6A--ENGINEERING EMPLOYMENT SURVEY

Area of Employment	1968 NER*	1981 NCEE*
Consulting	26%	4%
Industry	47	70
Education	9	6
Government	14	15
Other	4	5
Total	100%	100%

Job Responsibility	1968 NER*	1981 NCEE*
Management	30%	34%
Design	20	24
Research & Development	15	12
Construction	--	4
Construct. & Consulting	12	--
Manufacturing/Operation	10	13
Sales	6	--
Teaching	3	6
Other	4	7
Total	100%	100%

*NER = National Engineers Registered by NSF
        of 53,000 engineers
*NCEE = National Council of Engineering Exam-
        iners survey of 3879 registered engi-
        neers

It soon became apparent that four-year curricula
could not provide the education graduates needed today
for progressive practice during the ensuing half century of their active careers. The Goals Study (1968)
recognized the need for graduate doctoral study for

some, and a lengthening of the program for all others to a five-year master's as the first designated degree in engineering. Unfortunately, neither the colleges nor industry accepted the recommendation of the prestigious Goals committee. Students, however, were alert enough to recognize the logic of the need for more formal education for their future. This is evident from a study of Figure 6B which shows that M.S. and Ph.D. degrees awarded in engineering increased from 1952-82 until they now comprise about 30% and 5% respectively of B.S. degrees awarded. Actually, these figures should be considered as being about 45% and 7% of those B.S. graduates who remain in engineering, since about one-third transfer into business, law, medicine, etc.

Concern was expressed in the early 1970's about the overemphasis on research-oriented programs and the virtual lack of practice-oriented programs. The various schools of engineering are shown diagrammatically in Figures 6C, D and E (92,96,65). These programs suggested admitting students into professional programs or schools at the junior or senior level, although it was to retain control over the freshmen and sophomore preprofessional programs. The professional programs were to bifurcate into research-oriented (M.S. and Ph.D.) or practice-oriented (M. Engr. or D. Engr.) programs after the 4th year (Figures 6C, D) or 5th year (Figure 6E). The effort would have paralleled educational study in biological sciences, where research-oriented Ph.D. programs in such fields as biochemistry, anatomy, microbiology, etc. are retained in arts and science colleges; and where practice-oriented M.D., D.D.S., and D.V.M. programs in medicine, dentistry, and veterinary medicine are restricted to such professional schools of medicine or J.D. programs in law to professional schools of law.

Actually the bifurcation of graduate engineering education into research-oriented and practice-oriented doctoral programs, with both founded on a common cultural and scientific core, was first proposed in 1949 (106). Bifurcation was also proposed for undergraduate programs into professional-general and professional-scientific types in the Preliminary (Grinter Committee) Report of October 1953. The latter suggestion was rejected almost unanimously then by engineering colleges. Neither proposal suggested the creation of separate schools for the two types of programs. Many engineering colleges now offer both 5-year Master of Science and Master of Engineering Degrees. In 1980 about

133

fifteen offered the 6-year Engineer's Degree and the 7-year Doctor of Engineering Degree but they are not yet as popular as the research-oriented degrees. Fifty years ago engineer's degrees required only original practice-oriented theses suitable for publication in major journals, but no resident course work. A few colleges formerly awarded either the Ph.D. or D.Engr. Degree for identical work, leaving their selection to the student.

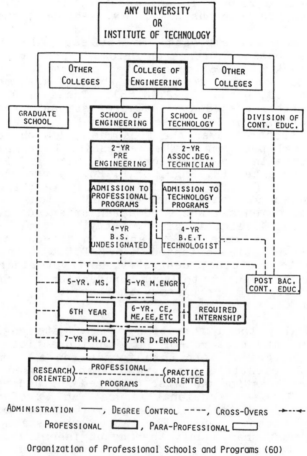

Organization of Professional Schools and Programs (60)
F I G U R E 6 D

134

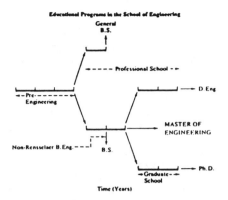

FIGURE 6E - EDUCATIONAL PROGRAMS IN THE SCHOOL OF
ENGINEERING AT RENSSELAER POLYTECHNIC INSTITUTE (189)

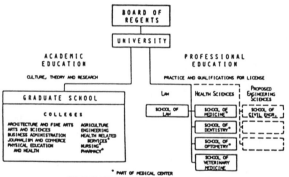

FIGURE 6F - RELATIONSHIPS BETWEEN A GRADUATE SCHOOL,
IT'S COLLEGES AND PROFESSIONAL SCHOOLS OF A
MODERN UNIVERSITY (60,126)

A 1958 ASCE survey (115) revealed that 66% of its
members favored 5-year curricula. Only 36% of deans
and 22% of CE department heads did at that time, but
both favored a core program that would be at least 50%
common for all engineering curricula (116). Then, at
the 1960 ASCE Conference on Civil Engineering Education, 30 eminent engineering educators adopted, virtu-

135

ally unanimously, the following resolution as the conference theme:

> Resolved that "the growth in universities and colleges of an undergraduate, preengineering, degree-eligible program for all engineers...with at least 75% interchangeable among various engineering curricula...be followed by a graduate CE curriculum leading to the first designated CE degree" (117).

Ninety-eight of the 144 CE departments in existence in 1961 responded to this resolution, and 59% approved it (118). The establishment of professional schools and of dual graduate programs of research-oriented (M.S. and Ph.D.) and practice-oriented (M.Engr. and D.Engr.) degrees was also favored by 76%. Despite the favorable acceptance of the proposals at that time, they were seldom implemented. Indeed, some undergraduate curricula drifted towards even greater specialization. However, in 1981, three state universities in Florida proposed both professional schools of civil engineering and a five and one-half year Master of Civil Engineering (MCE) degree. It also included instruction in professional ethics, law, communications, economics, management and 6-month internships or other "supervised clinical experience." The program was designed to meet criteria for ABET curriculum accreditation and NSPE professional school recognition.

Engineering programs may appear to change slowly, but they do change in time. The changes may be quite profound. For instance, the 1968 Goals Study recommendation of the 5-year M.S. degree as the first designated engineering degree would have made it the equivalent of the 4-year Central European Diplom Ingenieur (Dipl. Ing.) or the Dutch Ingenieur (Ir.) or the Scandinavian Civilingenior. Bjorhovde (107) cited the needs for the longer American program as being related to the more rigorous European senior high school studies. European high school graduates will have acquired the level of mathematics and physics, as well as the study of cultural subjects, not acquired by American high school graduates until after two or even three semesters of college level work. The Preliminary (Grinter) Report (1953) noted this weakness as follows:

> "There seems to be no major disagreement that an engineer cannot be trained to make effective use of modern knowledge of engi-

neering science in creative design within a
4-year undergraduate program.   It is even
more improbable that effective contributors
to research in engineering sciences whose development is now an accepted responsibility
of the engineering profession, can be trained
in four years...It seems more probable that
4-year training may be sufficient college
preparation for many students with general
professional objectives," i.e., for "production, construction, operation, sales, (equipment) installation, etc...The first four
years of the undergraduate program can hardly
be identical any longer for these two types
of engineering education which in this report
will be called professional-general and professional-scientific."

Although the engineering colleges refused to consider undergraduate bifurcation in 1953, and did decrease emphasis on design somewhat later, they did expand existing programs in 2-year technician and 4-year
technology programs until, in 1982, they were about
half as numerous as 4-year engineering baccalaureates,
as Table 6B indicates. "All of these 4-year programs"
fill an industrial need and "graduates may lead useful
lives. However not all have the academic base to practice on the high professional plane the future will require" (108) without further academic or self study in
technology, management and/or leadership.

TABLE 6B
ECPD (ABET) 1982 ACCREDITED PROGRAMS

Technology		Programs
2 Yr. Technician		445
4 Yr. Technologist		257
	Total	702
Engineering		
4 Yr. Basic		1220
5 Yr. Advanced		74
	Total	1294

Students did increase M.S. and Ph.D. engineering
graduate enrollments until 1970 (Figure 6B) when they
leveled off or began decreasing respectively. However,
the number of graduates enrolling in non-credit short
courses has expanded significantly.

137

Change has, therefore, been evolving, although the formal designated advanced engineering degree requirements remain elusive. This ill-defined nature of the engineering educational structure, and especially that of the profession as well, led Dean L. E. Grinter to regard the profession as an amorphous mass that refuses to congeal. Nevertheless, the unprecedented consequences of the technologies they generate is to impose change. It is imperative that engineering education change accordingly so that it will prepare graduates for a half century of future practice.

"Professional engineers, educated well enough to be eminently able to practice and adequately motivated to appreciate their heritage, will be better able than most to understand the impact of technology on society. Their contribution toward progress can be immense. Whether their education will prepare them for that challenge and such leadership will depend upon what educational changes educators and practitioners alike insist upon before the novices entrusted to their care are admitted to the profession." (109)

There are many reasons why four year programs cannot fulfill all of the objectives to which professonal schools of engineering might aspire. Time is but one deterrant. These objectives might be listed as follows:

1. To educate novices for the responsible practice of a specified professional art.
2. To transmit applicable existing knowledge after first "distilling" it for concise presentation.
3. To search for new knowledge that enhances the art involved.
4. To teach technological innovation and entrepreneurship.
5. To convey a sense of ethics and professionalism.
6. To motivate novices for public advocacy roles to protect its health, safety and welfare, and resources and environment.
7. To groom societal leaders for a technological civilization that will protect freedom.

"Educational programs should be long enough to fulfill their objectives. Today's engineering profession should also require its designated graduates to be cultured and technically literate. A six year program plus some internship might suffice now... Proctor enunciated that concept in 1939 when he was President of ASCE. He said,

"Is it not time we should agree that a professional man cannot be produced in four years, but that an accredited civil engineering training must be definitely post graduate, with a broad undergraduate training as a pre-requisite? (110)

This discussion does not imply that B.S. graduates will not make good taxpayers or employees. The discussion merely suggests that more education should be prescribed before the individual is designated as an engineer. Whether that extra education should be absorbed on campus, or off-campus through external degree programs, will be treated later" (60).

Actually, the profession has always considered the 4-year engineering graduates only as excellent potential, even though industry usually classified them as engineers when it employed them. The profession never regarded the graduates as engineers until after they had acquired several years of experience. Thus, engineering four-year programs can be regarded as an excellent foundation for practice, but only after suitable experience has been obtained can society afford to trust the judgement of these practitioners, and particularly when the projects the engineers might supervise involve millions of dollars as well as of public health and safety.

As well balanced as engineering baccalaureates programs are, they lack two essentials; i.e., sufficient instruction to enable graduates to practice on today's advanced technological plane without additional study and/or experience, and a rigorous introduction to those non-technical subjects which would enable them to assume engineering management and societal leadership roles. Their education must groom them for that responsibility. We professionals who determine the edu-

139

cational patterns engineering students must follow must always remember that, just as we live today in what was the future yesterday, society's future will likewise depend upon their abilities. The problem resolves itself into specifying how much formal education is enough to start, how such programs can educate both specialists and generalists, and how the education can best be continued lifelong.

The intimate relationship between education, leadership, culture and history was described ever so prophetically in 1940 by the late newspaper columnist Walter Lippmann (113). He predicted that prevailing education would, if continued, destroy civilization. Those "responsible for education have progressively removed from the curriculum studies of Western culture which produced the democratic state...sending out into the world men who no longer understand the creative principle of the society in which they must live...the moral order...the religious tradition...the conception of law...Those who are responsible for modern education are answerable for the results...They have educated the politicians, businessmen, professionals...and the educators!" (114).

SPECIALIZATION

Specialization of engineering education accompanied the splintering of the profession itself, which became apparent about 1870 when the fragmentation of the technical societies began (see Table 4A). The ability to compress all knowledge needed for engineering practice into a four-year program soon faded. Periodic efforts have been made to retain a core of essential basic and engineering science in all programs, but the core keeps shrinking. In fact some engineering programs now exclude any instruction in thermal or electrical science, and minimize the study of applied mechanics, of the structure of matter, and of its behavior when subjected to the various forms of energy. Similarly, the study of introductory subjects related to industrial operation, like economics, law and psychology; or to societal leadership, like ethics, history, and customs is barely required if, indeed, it is included at all.

It should be obvious that no educator can continue to squeeze all relevant knowledge into any program of finite length when knowledge is accumulating faster (Figure 2A) than he can distill it or exclude the obso-

140

lete. Nor can anyone master enough knowledge of more
than a few specialties to apply it effectively enough
to compete as a practitioner and not endanger the public safety. Many wonder why the engineering profession
continues to specify the 4-year program is its first
designated degree, when all other professions rely on
longer programs. One reason is that industry is usually so short of capable technical employees that it
makes formal post-baccalaureate study uneconomical by
offering high starting salaries. Another reason is
that the 4-year program is used to educate individuals
for a wide spectrum of technological employment. However, there is a danger that too few of these graduates
will absorb the experience or knowledge to become the
generalists our technical society needs, even though
enough barely do acquire what they need as specialists.
David Rose, a former director of the Oak Ridge National
Laboratory, commented on this educational dilemma as
follows:

> "Most university students do not spe-
> cialize in science or engineering. Their
> non-specialization is reasonable, but their
> almost total ignorance of the technological
> sub-world...is unreasonable...These issues
> are more easily resolved for engineering and
> applied-science education. Two opposing
> views contend: specialization first and gen-
> eralization first. Alas, the...systems...
> lean toward the former. Disciplinary bias,
> the belief that an engineer with even minimal
> training is good for something, and the fear
> that a generalist is useless, combine to pro-
> duce engineers who are relative technicians,
> less aware of social issues than the general
> public...
>
> The opposite approach...would be better.
> It is happening anyway...Four-year college
> education is...designed for popular consump-
> tion...many technology-based companies demand
> advanced degrees as the professional norm...
> it would make sense to shift engineering spe-
> cialization into graduate years. Undergradu-
> ate engineering could then become 'pre-engi-
> neering'...
>
> Why do civilizations decline?...We ima-
> gine we can solve class A problems with class
> B people...Trusting over much in organiza-

tion, we find ourselves gummed to death by a
horde of intellectually toothless bureaucratic mice, some of whom we trained" (111).

Increasing specialization leads to an unwilling-
ness to assume accountability or responsibilty--to pass
the buck (Article 2E)--and reinforces the concept that
engineering is a reactive profession. Robert Seamans,
a former president of the National Academy of Engineering, commented on this criticism and engineering education for the future as follows:

"This view of engineering as a reactive
profession, in which practitioners respond to
demands with their skills, ignore...the creative aspect, in which new developments precede perceived needs. A classical example is
the airplane...(T)he camera, the telephone,
and the phonograph (were also) creative, not
responsive...

Technological developments during World
War II suggested that less structured education produced more flexible and creative
technologists...Today (1981) engineering education is in a fluid state, with wide-spread
conviction that changes should be made.

Society will continue to need...engi-
neering scientists, practicing engineers, and
entrepreneurial engineers, and...In addition,
the systems engineer and the management engineer.

The most difficult...educational challenge involves...systems engineers; the technical community is concerned that programs
encompassing sociology, economics, and political science lack professionalism and are of
passing interest...(T)he most attractive
graduate program...related to entrepreneurship and management...is the professional engineer's degree that allows time to acquire
the needed breath and depth...(and) room for
a substantial thesis without research emphasis" (112).

How, then, are all of these requirements for generalists and for specialists, educated well enough to
assume corporate management, societal leadership and

entrepreneurship to be accomplished? It is obvious that it cannot be done with a 4-year span, nor with only one program. An English educator once suggested that schools should educate students and let the world classify them. But that might lead to a technological mismatch with people focusing on different time-scales and objectives to solve societal problems, or to an overeducated segment of unemployed graduates incapable even of enjoying leisure.

Society can profit by divergent trends provided those who lead it are educated broadly enough to follow the advice of Britain's late Prime Minister Disraeli. He suggested leadership that conserved what is good and changed what is not. "The scientific base on which professional practice must be based is changing constantly...But moral and ethical codes also form a base for professoinal practice. They also change with time- -and throughout the world as well" (8). These will be considered in greater detail in Chapter 7. It seems evident that different forms of supplemental education must be superimposed upon a basic foundation. One can only wonder, however, why the engineering profession in 1980 had seen fit to have accredited 20 different 2-year technican programs, 21 different 4-year technology programs, and 22 different 4-year engineering programs.

SUPPLEMENTAL PROGRAMS

Arguments continue as to what scientific and cultural subjects should be considered as basic, and of how much is enough. Perhaps the simplest plan is to suggest two criteria; 1) to insist upon enough specialization so that graduates can read and digest technical literature well enough to evaluate and apply it, and 2) to require enough generalization so that graduates can appreciate their cultural heritage while supervising interdisciplinary designs and exercising leadership in industrial and governmental arenas (8).

Programs that supplement the technical criterion include undergraduate cooperative industrial experience, graduate research assistantships, and graduate practitioner internships. Programs that augment the general criterion include service as Congressional or White House fellows; as assistants in the Office of Science and Technology or on the staffs of U.S. Congressmen, Senators and Congressional Committees; or as Fulbright graduate appointees, etc.

In my half century of teaching I found engineering
co-op students as a group to be the most motivated and
discerning. The extra year they spend in industry ex-
tends their baccalaureate program to 5 years, but industry employs them at essentially the same starting
salary it pays M.S. graduates without any experience.

Congressional fellowships were inaugurated and
sponsored in the late 1970's by engineering societies
with a dual purpose in mind: 1) to provide experience
for recent engineering graduates so that they could understand how societal technical missions are generated
in legislatures, and 2) to supply legislators with
knowledgeable assistants who could study technical
problems in depth and render appropriate advice.

Comparative supplemental programs for professional
education in law, medicine, architecture and engineering are illustrated in Figure 6G. The pre-professional
and professional educational requirements, entrance
criteria, degrees awarded, internship and/or apprenticeship expected and licensing available are detailed.
Two diagrams depict present and suggested future programs for engineering. The profession seems to be
moving toward dual programs for which the chart labeled
"present" represents current research-oriented education, and the lower chart labeled "future" represents
practice-oriented education. Both are available at
selected institutions, as Figures 6C, D and E indicate.

There were no "formal" professional schools of engineering in 1980 whose administrative structure within
their university paralleled the professional schools in
law and the health sciences. This does not mean that
existing schools of engineering are unprofessional, but
they were not established by separate state statutes as
was usually the case for law and medicine. The administrative structure of the several schools in academic
vs. professional education are illustrated in Figure
6F. It should be noted that existing research-oriented
engineering programs are retained in the graduate
school, but that proposed practice-oriented engineering
science programs for the several specialty disciplines
(civil, mechanical, etc.) would parallel those for the
medical sciences (89,126).

The foregoing paragraphs have described suggested
changes in engineering education that were adopted
slowly, if at all, by the profession. One could easily
argue that the 4-year program requirement for all engi-

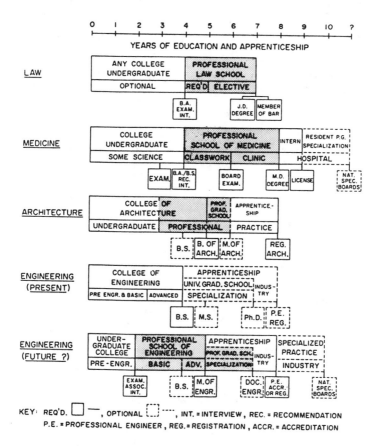

FIGURE 6G - COMPARATIVE PROGRAMS

OF PRACTICE-ORIENTED EDUCATION   (89)

neering students, plus voluntary supplemental study for
perhaps half of these is satisfying the free market,
judging by the starting salaries being paid. Such rea-
soning fails to include the cost to society of the ca-
tastrohpic losses associated with the Love Canal waste
disposal, the 3-Mile Island nuclear reactor and the DC-
10 crashes (Articles 4F and G), nor of the cost of the
regulatory agencies (Table 5A). The early specialization of engineering education failed to produce either
the specialists or the generalists to prevent these
failures.

145

Early baccalaureate specialization has likewise failed to provide a broad education for those students who elect to change curricula. Only one-half follow their undergraduate specialty, and one-third leave engineering for careers in management, sales, business and other fields as Table 6E shows. The 1970 Census reported 1,000,000 practicing "engineers" and 480,000 who had transferred out.

TABLE 6C-MOBILITY OF ENGINEERS DURING 1971-74
OKLAHOMA STATE SURVEY--1000 CASES   (119)

Original Specialty	Stayed In Specialty	Transferred Within Engr.	Left Engr. Practice*
Aeronautical	33%	15%	52%
Chemical	61%	23%	16%
Civil	56%	5%	39%
Electrical	58%	5%	37%
Industrial	47%	15%	38%
Mechanical	48%	21%	31%
Metallurgical	50%	32%	18%

*For Management, Sales, Business, etc.

Similarly, Figure 6H shows that the fewer years of formal education an engineer has, the younger he is, the more mobile he is likely to be (120,119).

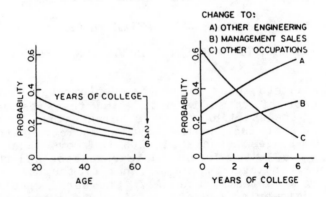

FIGURE 6H - PROBABILITY OF OCCUPATIONAL CHANGE
FOR CIVIL ENGINEERS (119)

Industry's preference for an educational status quo can also be understood by considering its needs for new graduate engineers and technologists. According to Bureau of Labor Statistics (Table 6D) these will grow from 98,000 in 1976 to 131,000 in 1985 (119). However, available graduates were estimated to increase only from 53,000 to 70,000. The manpower deficit must be made up from science majors, etc. to compensate for the 45% of technical employees who transfer out.

TABLE 6D--ENGINEERING MANPOWER REQUIREMENTS (119)

	New Engineers Available		New Jobs Available	
	1976	1985	1976	1985
Graduates/Yr.				
Engr. BS + BET	53,000	70,000		
Sci. Related	10,000	?		
Transfer In	30,000	?		
Job Openings/Yr.				
Growth			31,000	41,000
Deaths, retirees*			23,000	31,000
Transfer out*			44,000	59,000
Immigrants	5,000	?		
	98,000	?	98,000	131,000

*Variable estimates depending on market conditions, etc.

The need for supplemental programs to educate these transfers, as well as to provide continuing education for the engineering graduates is, thus evident. On-campus collegiate education has become too expensive to prolong beyond an absolute minimum time span and could be supplemented as the following quotation suggests:

"Post baccalaureate study need not necessarily be confined to any university campus...There is no need today to constrain advanced study to a classroom with a series of lectures. The advent of televised lectures, video tapes, cassettes, programmed lessons, etc., preclude that. Guided self-study programs, duly certified with appropriate degrees are just emerging in the United States. They have been available at least at one prestigious British university (London), which has for over 100 years granted "external" degrees on the bachelor, master and

147

doctoral levels without any residence requirement. Its external baccalaureate was
discontinued in 1979 because of the difficulty of obtaining laboratory experience. Its
MPhil and PhD degrees require only research
dissertations or theses but no course work.
Other external higher doctorates include MD
(medicine), DD (divinity), DLit (literature),
LLD (laws), DMus (music), DScEngr (engineering) and DScEcon (economics). These degrees
are decidedly harder to earn and are regarded
at least as highly as the normal "internal"
resident degrees...

American universities could consider inaugurating such "external" doctorates emphasizing design-related, practice-oriented programs and use the entire industrial complex
as part of their educational environment.
Industry today is well staffed with engineering practitioners who earned research-oriented degrees and learned afterward how to apply
their theoretical training to practical problems. The aid of these engineers could be
enlisted...The external programs suggested
here would in no way duplicate resident research-oriented study which now educates
graduates for 15 percent of the job market.
Instead, the programs would be broadened to
include those other elements necessary for
successful engineering practice, management
and leadership, without diminishing the advanced theoretical base needed for practice
at the highest technical level" (120).

There is an accelerating need to optimize continuing and supplemental education. So far, the first designated engineering degrees have been entirely resident
ones. If the future will require a masters--or ultimately a doctoral--degree for entry into the profession, it might be advisable for professional schools of
engineering to develop the two-track approach (Figure
6D) for resident and non-resident (external) types.

"Scores of courses are already being offered by various engineering societies and
private organizations to "advance the art"
and to satisfy public pressure imposed on
state licensing boards. One large privately
financed organization, The American Manage-

148

ment Association (AMA), already employs 7,500
lecturers for 2,000 programs. It has 75,800
participants annually and is a $50 million
enterprise. It believes that "50% of schools
of business administration would find it hard
to justify their existence 'from the practical point of view'." It suggests that "half
of the schools of business should withdraw
from graduate-level education" (121). Such a
private organization might develop in engineering if its professional schools fail to
offer the appropriate service the profession
will need and which the public will demand.
These schools may find that some private educational enterprise may usurp post-baccalaureate, practice-oriented engineering
education" (120).

PROFESSIONALISM

Considerable space was devoted so far in this Article (6F) to 1) the length of engineering programs, 2)
amount of specialization, and 3) types of supplemental
programs. That the purpose of this entire educational
effort should embrace the philosophical foundation and
public purpose of the engineering profession would seem
to be obvious. Both have been defined and described in
Chapters 1 and 2 but, regretably, they have seldom been
alluded to in engineering programs, even through the
doctorate. Sinclair, commented as follows on the recent decline of professionalism in engineering: (122)

"(A) definite decline in professionalism
is not hard to document...The BART episode in
San Francisco, the Three Mile Island accident, the DC-10 (crashes),...the collapsed
buildings..., the...automobile (recalls)...,
have all involved some degree of engineering
incompetence...Pressure on engineering societies to curb...incompetence...and enforce...
codes of ethics...is increasing...(T)he perceived ethical problems arise more from professional incompetence than from technical
incompetence...A new design of engineering
education to produce...true professionals is
required. The first task is to develop a
philosophy of engineering as a profession."

Whether this decline or lack of instruction in
professionalism was due to the absence of an engineer-

149

ing philosophy, or to early over-specialization in beginning 4-year programs, or to over-emphasis on faculty research and publication is difficult to prove. All probably contributed, but I believe most of the blame can be ascribed to early specialization and the consequent faculty enthusiasm for it. Both leave little time for inculcation of a professional attitude. Brown and Lynn, directors of Cornell's Program on Science, Technology and Society, commented on this dilemma as follows:

"...professionalism and specialization are not the same...professions, in order to be socially responsive and responsible, must...provide...specialists and generalists...American universities have provided an environment in which specialization flourishes and professionalism is demeaned in all fields...professional expertise...is founded upon...bodies of theory and knowledge that have great generality...It is virtually impossible to find any engineering problem that does not...intrude...into areas of social concern...Engineering needs to embody within every specialist the qualities of a professional...What is required is a change in the manner of teaching and the outlook of the faculty in every course...They must learn to discuss...social issues...to be...involved in providing liberal studies in a context in which they pertain to the education of engineers" (123).

The foregoing quotations cite the need for engineering education to include basic cultural studies so that practitioners can retain competence and develop adaptability in the ever-changing social environment their work produces. These studies should be integrated into design courses so that engineers will comprehend the complexity of the society in which they practice. Lynn cautions that faculty must accept responsibility of the total curriculum—and to "educate a person capable of understanding the moral and political consequences of his actions as a specialist" (124). Specialists, if they are to practice professionally, should not keep their "minds in a groove" as Alfred North Whitehead feared they might.

Perhaps what is most needed is a study of "history that explicitly deals with the role played by technolo-

150

gy" (124) so that engineers will be motivated to always consider the social impact of their own practice. That outlook will not be easy to teach, but it will pose a challenge worthy of a free society. Neither will a program encompassing professional attributes be inexpensive. The important point is that means must be provided for it if professionalism, and perhaps our free society, are to survive.

6G. THE COST OF EDUCATION

The existing and suggested engineering programs described in the foregoing paragraphs require significant funding by the colleges and investment by the student. Neither can be inexpensive if they are to be superior. The reader might ask, "Can society afford anything better?" Engineering education is expensive, perhaps even more so than medical education on a student-hour basis. The cost of laboratory equipment, so necessary for the student's "hands-on" experience that has been the hallmark of American engineering education, is exceedingly expensive. What is worse, the equipment becomes obsolete all too quickly.

One need only to contemplate the substantial cost and great need--and short life--of the vacuum tube-type analog computers of the 1960's. They were soon superceded by transistor-type computers that were cheaper to buy and to operate. Yet both imposed an absolutely necessary expense upon the engineering colleges that hoped to remain up-to-date. Similar cases could be cited for all of the other electronic and testing equipment that modernized college laboratories so quickly.

While the costs of operating engineering colleges escalated, so did the expenses of the students. If we are to provide the best education for these future professional leaders we might ask, "Can we afford anything less than the best?" The cost in money to society of engineering failures (Articles 3F and 4F) and of the governmental regulatory agencies (Table 5A) they generated--at least in part--were and are a significant percentage of the gross-national product. If the waste in resources associated with the recalls, shutdowns, collapses and crashes of these engineering products and systems were also considered, the cost escalates several fold. If these could have been prevented by our having educated superb professionals, society could have educated them, and supported the institutions

151

which trained them at a substantial profit! The complex technological society of the future cannot afford any but the best education possible for all of its professionals and its leaders.

The seriousness of the financial support available to engineering colleges and students was noted in a NSF/Department of Education Report prepared with ASEE and AAES assistance in response to President Carter's memo of February 1980 (125). He expressed concern about the adequacy and quality of engineering education and graduates for long-term needs. The Report cited the crisis in obsolete laboratory equipment; faculty shortages, industry/college salary differentials; insufficient graduate stipends; and deteriorating test scores in science, mathematics and verbal facility of high school students applying for admission to colleges. The Report recommended federal support to modernize laboratory research equipment, salaries, training and facilities; to provide graduate stipends; and to coordinate science teaching in high schools and colleges. Major industries cooperated within a year donating millions for salary supplements, equipment and graduate internships and fellowships. All of these grants in aid will help. However, it is doubtful whether a serious comparison has been made between the cost of educating engineering professionals in a first class way and the savings their practice would make to the overall economy.

The wealth lost to society from faulty designs, in addition to those mentioned above, is also illustrated in the following quotation:

"The recent structural failures of auditorium domes and storage dams have been adequately described and their loss estimated in engineering journals. Additional recent failures in Florida of drainage projects and municipal waste disposal systems have been estimated by Matheny (126) at 100 million dollars, or enough to fund the operation of a graduate professional school of civil engineering for 50 years!...A similar study of the design and operation of aircraft, nuclear power plants and automotive products would reveal even larger costs for other engineering specialties" (60).

152

The problem of how to finance the education of eminently qualified engineering practitioners still remains to be solved. Arrangements must be made with industries and government laboratories for the support of fellowships and related internships similar to the government stipends available for law and medical students, and for most undergraduate students through ROTC programs. These stipends require some subsequent governmental services.

"The cost of post-baccalaureate education, be it research- or practice-oriented, is expensive. Gloyna (127) showed that the 1976 cost of engineering education in Texas amounted to $17,900 for a B.S. degree, excluding about one-half for science and cultural courses, an additional $18,200 for an M.S. and an extra $74,100 for a doctoral degree. This cost to society did not include the students' living expenses or tuition. During their lifetime, they should expect to repay this total in taxes or save society an equivalent amount through technical innovation. Their better training in the practice-oriented education of a professional school should also save society vast sums by better designs and fewer failures.

The cost to society of designating inadequately educated B.S. graduates as engineers may far exceed the funds that would be needed to extend their programs for an additional two years plus some internships to insure eminently competent practice. State licensure now guarantees only minimum "adequate" standards. It protects the public only from obvious incompetents who would practice as consultants. The public can only be compensated for losses after they occur." (60)

The cost to society and to the students could be reduced substantially if new professional schools were to offer external, post-baccalaureate, degree programs through the doctorate. Dr. Clayton Dohrenwend, former Provost of Rennselaer Polytechnic Institute, suggested at the 1974 ASEE Annual Meeting, that such engineering doctorates might be administered nationally by ECPD. Such supervised programs would encourage continuing professional development of engineers scattered throughout the world. These new schools should also

153

feel obligated to provide streamlined, organized, life-long, continuing practice-oriented education to maintain the graduate's technical and professional competence (60).

Schooling all too frequently is masked as education. Guided self-study could be far more effective and less costly. Professions which require prolonged on-campus instruction force universities into positions which closely resemble the old robber barons on the Rhine. The schools are forced to exact coin--and valuable time--from those who strive toward a career (128).

Post baccalaureate students could be financed, as Batdorf points out (129), by lending them "money on favorable terms...rather than...giving them grants. This would force them to figure out for themselves whether the education is worth the cost, instead of bribing them into taking advanced degrees as in the past" (129,7). Such limited, insured loans could be guaranteed by the government and made available to students without the endorsement of co-signers if provisions were made that would 1) prohibit any cancellation resulting from possible subsequent bankruptcy proceedings, and 2) prescribe future repayment from the student's automatic payroll deductions.

The best--and perhaps least costly--way to finance first-rate graduate professional education is with full-time, on-campus programs which embrace coincidental or subsequent internships. Such programs get competent graduating engineering practitioners "on stream" faster. Loans and co-op stipends could assist the student. Tax incentives, governmental grants, and joint industry/college research ventures could help the schools. The longer the educational programs are prolonged, the fewer are the students who finish and the longer it takes them to become fully productive.

> "The need for prescribed post-baccalaureate study and post-collegiate internship should be obvious. It has been to physicians for almost a century. It has been urgently recommended for lawyers by Chief Justice Warren Burger who believes that half of U.S. trial lawyers are unqualified to represent clients.

> Adequate study by professionals of our cultural and economic heritage will be needed

154

if our capitalistic system and political freedom are to survive. Tribus feels that "the U.S. and the free enterprise system face a monumental challenge...to prove that larger scale, urban technological society is possible..." (130). Engineering professionals must embrace adequate cultural education if they aspire to leadership in the industrial, governmental or political arenas. Society has nowhere else to turn for political leadership but to those professionals it still trusts. Too many politicians do not lead; they are led by lobbyists." (60)

6H. PROGRAM ACCREDITATION AND RECOGNITION

The study of the foregoing paragrpahs might lead the reader to wonder why it takes so long to change policies and procedures. The libraries are full of articles suggesting ways to improve engineering education, and one society (ASEE) has devoted its efforts to that goal since its founding in 1894. The situation could be compared to the one which the British statesman, Cecil Rhodes (1853-1902), faced in South Africa. Years later he was quoted in "Last Words" as having said, "So much to do; so little done."

A great deal of progress has been made during the last half century as this chapter has noted. One of the most significant additions to the educational effort occurred in 1934 when the Engineers Council for Professional Development (ECPD) was created, mainly to accredit engineering programs. Its name was changed in 1980 to the Accreditation Board for Engineering and Technology (ABET). It differs significantly from university accrediting associations which accredit whole universities and all colleges within their structure. ECPD, now ABET, accredits only individual engineering curricula. Hence, its policy avoids blanket approval of strong and weak programs, some of which may not meet its demanding criteria.

Its criteria include standards for cultural and scientific subject matter, engineering analysis and synthesis, faculty competence, laboratory equipment, financial resources, library facilities, etc. Table 6B lists the types and numbers of such programs accredited in 1982. Undergraduate "specialties" accredited in engineering number 22. The master's level (practice-oriented) accreditations number about 6% of the baccalaur-

155

eate programs. Not all engineering colleges offer graduate work. However, the majority of engineering colleges which do elected not to apply for such accreditation, fearing it would prevent educational experimentation.

If each specialty that comes--like genetic or biomedical engineering, or goes--like railroad engineering, is to be accredited, one could easily see that constraints would hamper the development of new programs. One must wonder, however, why every new specialty must be approved, and why these could not be nurtured within existing bona fide schools of engineering. Other professions like law and medicine accredit schools, not individual specialties like corporate law or pediatrics, or whole universities of which the schools are a part.

Accreditation is but one of the many pressures exerted on professional education, the primary objective of which is to protect the public. Most professions other than engineering have found it necessary to accredit their professional graduate schools. As was noted above, ABET accreditation was extended to include advanced work of a professional nature, but not programs which are solely research-oriented. Extensions of this effort have been suggested, as the following quotation indicates:

"The nature of the present and possible future accreditational requirements, beginning after two years of pre-engineering, are shown in Figure 6I. This chart also shows how the present research-oriented programs with suitable crossovers could be retained in university graduate schools, leaving the practitioner-oriented programs to professional schools,...Extensions of advanced programs beyond the master's degree to include a professional degree in some specialty, or subsequently the doctor of engineering degree, are also shown." (89).

6I. EDUCATIONAL REQUIREMENTS FOR LICENSURE

Engineering Registration Laws in all states and territories of the United States require registration of all engineers who offer their services to the public, if public health and safety are involved, unless they are exempted. Most laws exempt those engineers

156

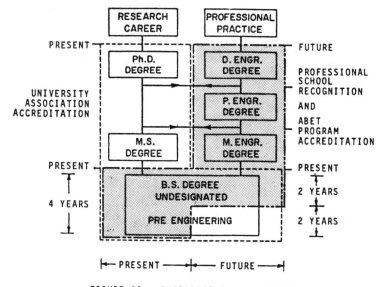

FIGURE 6I - ENGINEERING EDUCATIONAL
AND ACCREDITATIONAL REQUIREMENTS (89)

who are employed by any governmental agency, any public
utility (electric, railroad, telephone, etc.), or any
industry engaged in interstate commerce, or who are supervised by a registered engineer. Perhaps not more
than 5-10% of all engineers are, therefore, required to
be registered. Many more are; their number is about
40% of the total.

All states now generally require graduation from
an accredited 4-year engineering curriculum in order to
be permitted to take the first 8-hour examination--that
on Engineering Fundamentals. Engineer-in-Training
(EIT) certificates are awarded to those who pass. They
are allowed to take the second 8-hour examination on
Professional Practice and Principles after having obtained four years of experience acceptable to the board
and authorized to practice if they pass.

Graduates of some related science curricula or
non-accredited engineering programs are also allowed to
take the examinations but usually only after they have
obtained two years of additional experience before being admitted to the EIT examination. Graduate study in

157

engineering is accepted as experience with one-year being allowed for a masters and an additional year for a doctor's degree. Table 4B illustrated how the education/experience ratio varied for six professions. Similarly, Figure 6G outlined the educational and internship requirements for licensure in law, medicine, architecture and engineering. It also illustrated criteria for national specialty board status for present day medicine and possibly for engineering of the future.

There are various methods by which professionals can become qualified for practice besides the registration procedure just described. Article 4F listed these as licensure, registration, certification and credentialization. Actually, engineering registration is a licensing procedure by which the state keeps a register of those it has authorized to practice.

Until recently licenses were granted for life. Most states began in the 1970's to require evidence that the practitioner's knowledge was not outmoded. Admissible activities included completion of graduate courses, participation in seminars, attendance at professional meetings, etc. Medical and legal requirements for relicensure are more demanding than are those in engineering, mainly because these professions began the recertifying process earlier.

The years in which several of the principal professions in the United States first instituted licensure, civilian collegiate education, program or professional school accreditation, etc. is shown in Table 6E. Formal professional curricula in military engineering were begun at the U.S. Military Academy at West Point, N.Y. in 1817 (Article 6D), and in non-military (civil) engineering in 1819 at Norwich University. Table 6E indicates "That engineering has lagged behind its sister professions in conceiving an integrated professional program, in accrediting curricula, in passing licensing laws, in establishing a National Board of Examiners, and in founding a professional society. Omitting the first (state) licensing law and first college course, this lag varies from 28 to 87 years and averages about 48 years" (117). Table 6E does not indicate when National Specialty Boards were first organized for medicine. No other professions had seen fit to do so up to 1982. Specialty boards and methods of qualification for practice were treated previously in Article 4F and Figure 6G for medicine and engineering. The future may show that engineering qualification need not be

158

restricted to formal licensure, but that continuing education and perhaps future specialty board status will be required for engineering professionals.

TABLE 6E - COMPARATIVE PROFESSIONAL DATA (117)

	Medicine	Dentistry	Law	Engr.
First Licensing Law	1760	1841	1980	1907
First College Course	1765 (U.Pa)	1840 (Balt)	1799 (W&M)	1819 (Norwich)
2-4 Plan Adopted*	1918	1937	1921	19??
Prof. Society Accreditation	1907	1909	1921	1936
National Board of Examiners	1892	1883	1931	1920
Founding of Professional Soc.	1847 (AMA)	1859 (ADA)	1878 (ABA)	1934 (NSPE)

*2 Yr Pre-Professional followed by a 4 Yr Professional Program

It is interesting to note that neither medical, law nor engineering schools require licensure for their faculty (65,92).

"Professors of preclinical subjects like anatomy, microbiology, and biochemistry frequently are not licensed to practice, and may hold Ph.D. rather than M.D. degrees. Medical colleges do not insist that such staff members practice medicine to teach this preclinical material. Medical colleges do insist that those faculty teaching the clinical--or practice-oriented--courses practice medicine by "laying their hands" on patients so that their students can learn the art adequately. Typical practice-oriented courses are surgery, gynecology, pediatrics, etc. The colleges do not insist that these professors be licensed but to require that they be graduates of accredited medical schools. But the State Medical Licensing Boards do insist that all physicians who practice by treating patients be licensed...

Law schools employ a procedure analogous to medical schools in appointing faculty. In general, the professors must be lawyers;

159

i.e., graduates of accredited law schools, but are not required to be members of the bar (which gives them the right to practice in court). However, the chance of obtaining a faculty appointment without having had previous trial experience is probably small. Nevertheless, Virginia law grants law professors of its state schools who have taught full time for three years the privilege to petition the court to be admitted to the bar without examination...

The practice of law and medicine is restricted entirely to those legally qualified by licensure, whereas most engineers can escape the need for legal licensure because of exemptions and limitations in most state laws..."The question of whether engineering faculty should engage in consulting and if they should be licensed, whether or not they do consult, is disarmingly simple. But there is no simple answer...That policy would, as a condition of accreditation, obligate engineering schools to require those faculty teaching practice-oriented, design-related courses to engage simultaneously in consulting practice to keep abreast of the art and to improve it. The college policy need not require licensure. If the practice involves public health or safety, licensure by state registration boards will be necessary automatically" (65).

The strategy suggested in this quotation does not demand, or even suggest, that any or all engineering faculty be registered. Instead, the strategy suggests that only those faculty engaged in teaching "engineering practice" also engage in consulting. In so doing, they would keep abreast of their "art" and could also contribute to its improvement. The introduction a few decades ago of finite element analysis to industry is an example of such benefical reciprocity in engineering design.

7
Ethics vs. Licensure:
The Law and the Public Interest

7A. THE NEED FOR AND THE PHILOSOPHY OF
ETHICAL CODES AND LAWS

All civilized societies have found it necessary to
develop acceptable habits and customs in order to survive anarchy. Unacceptable customs had to be prohibited by law or by religion if these customs seriously
impaired the ability of the existing power structures
to function. Less serious violations were punished by
ostracising the offender at least socially by his peer
group.

Customs were ultimately divided into moral or ethical issues. Morals are principles or standards of
right and wrong conduct involving voluntary action.
Ethics may be considered as more of a study of human
actions as being right or wrong. Usually the issue is
not clear cut. For instance, one of the Ten Commandments, "Thou shalt not kill" is an unambiguous standard
for murder--except when the killing is in self defense.
Then the wrong becomes acceptable if not right. The
constraint against killing involves both legal and religious prohibitions and prescribed punishments.

Most other constraints are far less clear cut.
For instance, the Commandment, "Thou shalt not covet,"
is not illegal but it is immoral. Although the action
here is entirely voluntary and not reflexive, as is the
case for self defense, the prescribed punishment is
virtually non-existent. It may involve nothing more
than social censure.

These preliminary remarks may help to explain why
professional codes of ethics were so long in developing, and why they are enforced so feebly. Ethics provides a way of probing the acceptability of customs. I
prefer to associate moral standards with religious issues as they apply to the acceptable behavior of indi-

161

viduals, and to associate ethical standards with secular issues as they apply to the recognized objectives of professions. Both moral and ethical standards are necessary for a stable society.

The difference between ethics and law was described by Earl Warren, a former Chief Justice of the United States Supreme Court (131,132), as follows:

> "Society would come to grief without Ethics, which is unenforceable in the Courts, and cannot be made part of Law...Not only does Law in a civilized society presuppose ethical commitment; it presupposes the existence of a broad area of human conduct controlled only by ethical norms and not subject to Law at all...
>
> The individual citizen may engage in practices which, on the advice of counsel, he believes strictly within the letter of the Law, but which he also knows from his own conscience are outside the bounds of propriety and the right. Thus, when he engages in such practices, he does so not at his own peril--as when he violates the Law--but at peril to the structure of civilization, involving greater stakes than any possible peril to himself...
>
> This Law beyond the Law, as distinct from Law, is the creation of civilization and is indispensable to it..."

If all moral and ethical standards were transformed into laws, the burden of enforcing them would be overwhelming and the court would become hopelessly entangled in legal trivia. The legal system might well collapse. There is, thus, a great need for ethical systems which are extralegal.

There is a danger to society, however, if the ethical standards are neither widely known nor followed. Like their moral equivalents, ethical standards involve voluntary acceptance; like legal equivalents, their violations should not be ignored. Ideally, charges should be filed against violators, hearings scheduled, and punishment prescribed if the charges are proven. Such actions, however, can only be taken if the ethical standards are formulated into accepted codes by such

162

peer groups as professions, and then, only if the individual professionals agree to be bound by the code's constraints. Passive codes, which are merely suggestive and involve no penalty if they are broken, probably do little good.

The learned professions usually require applicants, who apply for membership in their societies, to agree to be bound by the constitution, bylaws and code of ethics of these organizations. Unfortunately, as we shall see in Article 7E, only a few of the major engineering societies have codes or enforce them effectively. Thus, even mandatory codes are far from perfect.

"Codes of ethics are, as Ivan Hill points out" in THE ETHICAL BASIS OF ECONOMIC FREEDOM (44) "covenants between a group of peers and the public. Unity of the peer group is essential in order to define acceptable professional behavior. Barzun has noted that an ethical 'code only sets the limit beyond which behavior is condemned' (21). If it is condemned, the public expects that action will be taken by the group against the violator. Ethical codes of groups upon which the public has conferred professional status cannot be mere suggestions for ideal behavior like most moral codes. The late newspaper columnist, Walter Lippman, has said that an authoritative code of morals has force only when it expresses the settled customs of a stable society...

The public, however, will require more than a professional's awareness of ethics as he practices. It will insist on ethical code enforcement or it will relegate the profession to vocational status and replace its advocacy function with a (governmental) regulatory body. If the profession is to survive and to enforce its code, it must achieve unity of practicing individuals more so than of engineering societies. This, engineering has yet to do" (60).

Engineers have not united into an umbrella organization of individuals probably because their undergraduate curricula were splintered into a dozen branches a century ago. They think of themselves as specialists first, and then as engineers. Likewise they have not

163

considered ethical codes as absolutely necessary after their graduation perhaps because they were seldom taught much about the need or function of codes in college. In fact, only about half of those who do graduate and remain in engineering even bother to join any engineering society. It might be easiest to accomplish a genuine regard of the public's need for professional codes by having all accredited engineering curricula emphasize such instruction. This could be accomplished readily through guided self-study for all students and by subsequent examinations in this topic as a prerequisite for any engineering degree. Such an accreditation requirement extended over a decade might be enough to achieve that goal since each year's graduates add over 5% to the total number practicing.

The foregoing remarks do not imply that engineers generally are unethical. In fact, Gallup polls since 1974 have ranked only clergymen and physicians ahead of engineers in honesty and ethical standards among the 20 occupations surveyed. Yet, among the million or more practicing engineers there have been enough who succumbed to managerial pressure and permitted unsafe products or systems to be produced. The resultant waste in resources was serious enough, but it was the impairment in public health and safety that led to the creation of ever more governmental regulations. Had the engineering profession been unified enough to hold these few violators *accountable*, the losses would have been far less. However, if individual practitioners are to be held accountable as they uphold their profession's code when public health and safety are involved, they should be supported by their professional societies. These ombudsmen functions (Article 4G) would then allow the dialog to occur between equals; i.e., the engineer's employer and his professional society, rather than between him alone and his employer. It seems reasonable to suppose that discussions meriting such dialog would profit from the added confidential counsel, that equitable solutions could be developed in most worthy cases, and that the corporation, the public and the the profession would all benefit.

The foregoing remarks assume that ethical codes, like professions, do have a public purpose. If the codes merely specify those minimum moral standards which condemn lying, cheating and failure to keep a promise and which apply to all rational people, there would be no need to reiterate these criteria again. But the ethical code of the engineering profession is

164

linked directly to its fundamental purpose to protect the public's health and safety, to conserve its resources, and to preserve its environment.

Ladensen (133) questions whether codes which do have a genuine basis in morality should be enforced because "punishable moral transgressions constitute only a small subset or morally unjustifiable behavior...we have made killing punishable. But we don't...do likewise with...lying, cheating, or breaking promises." He cautions that "professional people posses expert knowledge which often enables them alone to identify unethical behavior by their peers." This strongly suggests that they should have the responsibility of imposing sanctions and of identifying violators.

Most people are reluctant to report crimes, let alone unethical behavior, and the immense size of most professions results in natural indifference to individual peer behavior. Yet the Guide to Practice used to interpret engineering codes of ethics (APPENDIX B) stipulates that "Engineers who have knowledge or reason to believe that another person or firm may be in violation of any provisions of Canon 1* shall present such information to the proper authority in writing..." (Guideline 1d). Should he fail to do so, he himself violates the code. I know of no case where such omissions have ever been even investigated by professional ethics committees. "The problem of encouraging members of a profession to ferret out violators requires counteracting...the most deeply entrenched social tendency of our time...Certainly a return to the frankpledge system of the middle ages, whereby...responsibility for the acts of one's peers was engendered by holding everyone jointly liable...can be ruled out as a possibility." (133). So too can the probability that all professionals will become stoics and obey all rules freely.

Luckily for engineers, it is not so much unethical peer behavior as it is the infrequent questionable corporate production or operation relating to unsafe public use that is involved. If those involved; i.e., the engineers, their societies, the industries and the gov-

---

*Canon 1 (ABET Code of Ethics). "Engineers shall hold paramount the safety, health and welfare of the public in the performance of their professional duties."

165

ernment, all cooperate to resolve these issues before they become public and result in staggering economic losses there might be a genuine acceptance of ethical codes and of their enforcement. It certainly makes more sense to redesign potentially faulty products before they are marketed than to recall them afterwards, or to dispose of toxic wastes properly than to be forced to do so after whole watersheds have been poisoned. This assumes that the cost-benefit analysis is favorable, that the risk is minimized, and that the current engineering practice is within the state of the art. Otherwise the expense of subsequent legal fees, judgements, added regulatory controls, retrofitting and possible loss of life may far exceed the original economic gamble.

The need for laws is all too obvious. Dostoyevsky (1821-81) noted that, "If there is no God, everything is permitted." Society would then be completely anarchic and probably uncivilized. Society has elected to prohibit certain customs legally and to punish violators. Even if everyone were a saintly libertarian, who always abided by the golden role in dealings with his fellow men, there would still be need for laws that dealt with property rights and professional licensure. Civilization has grown so complex that no one has the knowledge to always select experts competent enough to serve him without, at times, endangering the public health, safety and welfare. Hence, society has found it necessary to license--or certify--innumerable vocations and professions like plumbing, cosmetology, engineering, accounting, medicine, law, etc.

Arguments have been made against all forms of licensure for a free society in which only the market place strategy, *Let the buyer beware,* alone would govern the purchase of all commodities and services. That strategy failed for two reasons; 1) there was greed enough to motivate a few vendors of goods and services to cut corners legally even if the public were endangered incidentally, and 2) a growing percentage of people preferred to let their government protect them against all hazards. As a result bureaucracies grew, regulations multiplied and freedom waned.

The difficulty involves the search for some reasonable compromise by which regulations are minimized and services are maximized. So far society has seen fit to license only those professionals whose practice serves the public directly; i.e., physicians, lawyers,

166

engineering consultants, etc. These practitioners are held legally accountable and financially responsible for their services. This public need was discussed more fully in Articles 2G, 3A and 4E.

7B. THE HISTORY AND PURPOSE OF ETHICAL PRACTICES AND CODES

The Hippocratic Oath of the medical profession is probably the oldest and most familiar written statement of professional ethics. It has been traced to Egyptian papyri of 2000 B.C. The Greek Hippocratic Collection was assembled about 400 B.C. but the Oath in its present form originated about 300 A.D. Since most professional organizations in the United States were founded in the late 1800's (Table 6E), they did not begin to adopt codes of ethics until after 1900 as Table 7A shows.

TABLE 7A - MAJOR PROFESSIONAL SOCIETIES

Professional Society	Founded	Code Adopted
American Medical Association	1847	1912
American Institute of Architects	1857	1909
American Bar Association	1878	1908
American Society of Civil Engineers	1852	1914
American Society of Mechanical Engineers	1880	1914
American Institute of Electrical Engineers	1884	1912
American Institute of Chemical Engineers	1908	1963
American Institute of Consulting Engineers	1910	1911

The early engineering ethical codes reflected appropriate conduct between peers more so than the profession's public purpose of protecting its health and safety. The earliest efforts date back to 1911-14 when the then young engineering societies adopted codes to govern their member's behavior; i.e., "how to relate to their clients" or "aimed...at avoiding competition among engineers" and "barred members from advertising, criticizing another engineer's work, underbidding rivals, trying to supplant them in their jobs, or from

167

acting 'otherwise than as a faithful agent or trustee'"
(134). Much of the emphasis on advertising and avoiding competition was declared illegal by the U.S. Supreme Court in 1978 (135) as being in conflict with the
Sherman antitrust law. These references have now been
deleted.

Well before 1978, codes like the Canons of the Engineer's Council for Professional Development (ECPD)
noted that the duty of engineers was to use their
"knowledge and skill for the enhancement of human welfare" and to "hold paramount the safety, health and
welfare of the public" (1974). Earlier versions of the
ECPD Canons stressed the need for the engineer to "use
his knowledge for the advancement of human welfare"
(1964), or emphasized "fidelity to the public" and the
"benefit of mankind" (1947), or "the obligation to
serve humanity with complete sincerity" (1943). However, the American Institute of Electrical Engineers,
in 1912, was the first to take the "radical step of
telling its members that their duty was to serve the
public interest" (134). It is this emphasis on public
advocacy that induces society to confer professional
status upon any group which endorses such service.

The gradual shift in emphasis of the codes from
personal behavior within the peer group to peer advocacy of the public interest developed for several reasons. Perhaps the most important was the change in the
engineer's employment status from consultant or owner
to employee over the past 200 years. By 1980, 90% of
the engineers were employees, yet 34% spent the bulk of
their time in management functions (Table 6A). A century or so earlier most engineers were consultants, educators or owners of industries. Names like John
Roebling, James Eads, Sylvanus Thayer, William Rankine,
Squire Whipple and George Westinghouse come to mind.
Other reasons involved the shift from complete management autonomy through the formation of unions to collective bargaining, then to the group rights era of the
1960's and 1970's concerned with minorities, and finally to individual whistle blowers involved in expressions of dissent. This change in the social structure
from one of a self-disciplined to a permissive one may
have been associated with the beneficial effect of
technology on living standards as well as on the environment.

Ethical practices were also changing because many
of the State Boards of Architects, Professional Engi-

168

neers and Land Surveyors altered the Rules and Regulations governing the licensing procedure, mainly after public pressure forced the appointment of a few laypersons to these boards. For instance, Virginia's Board declared them as "learned professions. (E)ach practitioner...is to be held accountable to the State and to the public by high professional standards in keeping with the ethics and practices of other learned professions..." Its Rules and Regulations for Professional Practice and Conduct include sections on responsibility to the public, public statements, conflicts of interest, solicitation of work, improper conduct, competency for assignments and approval of an unlicensed person's work.

Ethical constraints, thus, are beginning to infringe on the legal criteria for licensing of those practitioners, such as consultants, who offer their services directly to the public. Consultants comprise no more than 10% of the total number of engineers, for most state boards exempt from licensure those engineers employed by governments, public utilities, or industries engaged in interstate commerce. If, then, the remaining 90% are to be held accountable to ethical practices, it will have to be through engineering society membership. But who is to hold accountable those engineers who are neither licensed nor members of engineering societies? They constitute perhaps one-half of the total so classified by the census as was noted in Article 4C and 7A.

Whether society should license practitioners of selected learned professions, or even hold them ethically accountable is a question that has been debated persistently (Articles 2A and 4E). Engineers should, perhaps because they understand technology and are inclined to rely more on logic than emotions, participate in establishing societal goals. But, as Florman notes,

> "engineers have neither the power nor the right to plan social change...engineers are no more agreed upon how to organize the world than are politicians, novelists, dentists, or philosophers. Should we make small cars or large?...Accept the hazards of pesticides in order to feed hungry people? These are political questions...

The new ethics does not stop with considerations of public safety, but goes on to hold the engineer accountable for the quality of life...Engineers used to say that since they did things that society commissioned,... they could not be held personally responsible for any adverse consequences...this is essentially true...Not wanting to be taunted for being mere cogs in the social machine, and enjoying the feeling of importance that comes from being called shapers of culture, engineers have agreed to don messianic robes...

Political conflicts and philosophical debaters, however, cannot be papered over by appeals to engineering ethics." (136)

Florman stresses the need for ethicists to distinguish between creators and guardians, and not to confuse the functions of solving problems and establishing goals. It seems reasonable to me to consider the engineer as the creator of new products and systems as well as the guardian of our resources and environment. He should remain conscious of those professional obligations as he solves the problems he originates and the ones assigned to him.

7C. ETHICS AND PUBLIC ADVOCATES[a]

Politics and economics have always been intimately related. So have morals and ethics, ethics and law, and law and freedom. Freedom must be coupled with ethics in a capitalistic society if it is to remain uncollectivized. The relationship between freedom and politics borders on instability. "Stability is indispensible" for world order, as Arnold Toynbee once remarked, but "freedom is expendable."

If the engineering profession is to serve society, as distinct from the State, so as to maximize freedom and creature comforts as well, its professionals will have to operate in a complex social order with the constraints noted above balanced in proper proportion. Ethics and law will have to be coupled by optimizing voluntary ethical behavior and minimizing legal constraints. More laws will lead only to more authoritarianism and less freedom.

---

[a]Partial summary of sections of Reference 62 and 109.

Voluntary ethical behavior of engineering professionals may be maximized worldwide or nationally most readily by the organized effort of peer-group action. Peer groups can supply the needed moral and staff support for individual professionals trying to minimize unsafe or wasteful practices and for industries or governmental agencies endeavoring to maximize efficient resource utilization. A few professional societies now provide that help nationally. No such international organizations exist at all. The prevailing growth of transnational corporate empires will soon require the help of such global professional associations if these industries ever hope to reduce the various consumer regulations that now plague them in every nation where they operate.

The need to promote and enforce codes governing the operation of transnational enterprises (TNEs) is described in Ivan Hill's book, THE ETHICAL BASIS OF ECONOMIC FREEDOM (44). Such codes are needed for international investment and technology transfer if the TNEs are to earn their freedom from local governmental control. But these codes must be a covenant between TNEs themselves (as peers) and the several publics they serve.

Such codes can remain a substitute for national legislation, and extend economic as well as political freedom only so long as the vast majority of TNEs adhere to these codes and discipline those that do not. We can "be honest and free, or dishonest and policed" for "in the long run, honesty is the only...policy... compatible with a free market. Individually, this requires...the willingness to pay as much attention to corporate responsibility as to corporate profits" (44).

There will be occasions when professionals will have to help maximize corporate societal responsibility rather than profits, if they are to avoid generating still other governmental bureaucracies. The need for bureaucracies developed in the United States partly because a few professionals in a few industries permitted unsafe working conditions or products to develop by "cutting corners too closely." Now all industries are saddled with still more Federal regulations. The red tape generated by these agencies is now affecting all industries and universities to such an extent that President Derek Bok of Harvard suggested shredding all such useless regulations and using them to stuff sofas.

171

The need for these agencies here and abroad could have been avoided had the professionals involved in private industry been educated to always consider the long term impact of their practice on the public. Education alone for such ethical behavior is hardly enough muscle. Unger, in his book CONTROLLING TECHNOLOGY: ETHICS AND THE RESPONSIBLE ENGINEER (185), and others (53,54,55) have described cases where engineers who have pointed out unsafe conditions or designs to their managements have been fired for their pains.

The individual professionals must be aware of and dedicated to ethical precepts. But ethics has always been related to risk taking; with risks increasing as the population served, and the capital involved, decreases. Now, with resources shrinking, there is a danger of degenerating to life boat ethics; i.e., survival of the fittest nations (137). Nevertheless, the professionals involved, if they are to abide by and enforce any ethical code, and help establish public priorities, will need the backing of peer-group organizations and leadership at times when they deem it necessary to confront shortsighted management. Likewise, corporate managements and governmental agencies might also find it worthwhile to appeal to these same peer-group ombudsmen when technological-social issues need support. There was a beginning of such cooperation in the 1976 elections in the United States when referendums in six states rejected restrictions on nuclear power. Hopefully, such peer-group action will benefit both management and society. Worldwide professional organizations capable of lending such assistance were described in Article 4D (60,181) along with some details of their operation and organization.

The self-enforceable codes mentioned above are more likely to become realities if, as Yoder quotes Michael Pertschuk of the Federal Trade Commission, the mystiques which obscure professional practice dissolve so that

> "competition (can) probably supplant ethical self-restraint as society's safeguard against professional fraud...A community of self-policing people offering reciprocal services is seemlier than the atomized... marketplace...Professionalism has relied traditionally on the principle of self-regulation. One must always ask what the alternative is. And the alternative frequently

urged by current reformers is intrusive government regulation, or vigilante action by consumerist zealots or both...Tocqueville considered 'voluntary association' part of the essential genius of American democracy, and professional self-policing is an aspect of voluntary association" (138).

The idealistic principles of ethical codes frequently run head-on into practical situations. Engineers may then differ on which rules of conduct to abide by or to enforce, or even be unaware of them. Hall (134), for instance, reports that the results of a magazine survey in CHEMICAL ENGINEERING, in which readers were asked what an engineer should do when tempted to tell his new employer about a secret, unpatented idea used by his previous employer. Almost 71% of the respondents agreed that (the engineer) could divulge the information, provided it was not used for a competitive process. Not one respondent consulted the AIChE ethics code which prohibits such behavior.

Professional behavior constrains the engineer to so guide his practice that it remains within the ethical codes and legal bounds of the social structure in which he operates. Laws and codes change with time, just as do the technical innovations of his practice. Both laws and codes may also change with location as the engineer's practice becomes worldwide and international. Studies (109) show that ethics are slighted as competition increases and as enforcement decreases. Sometimes, with the mass of humanity, the only workable code of ethics is "anything goes." Professionals, however, should not seek that least common denominator nor favor the current status quo if it may get worse. Medicine's shift from private practice to clinics to socialized care is an example. Ethical codes, like professional curricula, may be geared too rigidly to the past and present and thus "create a false message of the future" (141).

Three ways of protecting the public interest have been presented here for engineering professionals to practice their art worldwide. They can remain *passive* and allow ever increasing governmental regulations to constrain their efforts if a few of their peers or employers endanger the public safety. Or they can maintain *active* individual ethical behavior and resign if they cannot prevent questionable practices. Or, they can *cooperate* with voluntary professional associations

173

for the self-regulation of their peers, and for the creation of safe technological products and systems. I believe that last choice is preferable and more democratic. It relies on individual ethical behavior and aggregate cooperative effort between industry, government and the engineering profession.

If this last, or cooperative, way is chosen, all engineers who claim that title would have to be members of some unified professional organization or be licensed by government to practice. Only then could they be held accountable by their peers. It is doubtful that universal engineering licensure will ever materialize. Industry could, however, insist that all engineers whom they employ and allow to practice be members of their professional organization. Again, I would prefer the latter. But, for it to become a reality, engineering education will have to be modified enough to instill an esprit de corps in the students it grooms for such dedicated service.

7D. TYPES OF ETHICAL CODES AND CREEDS[b]

The various types of the innumerable ethical codes in existence mitigate against their universal acceptance or application. Most professions, many of their associations, as well as business organizations have adopted ethical codes. All are multipurpose in nature and, thus, tend to be too long to remember. The 1977 ECPD (now ABET) Code with its four fundamental principles and seven fundamental canons is one of the shorter ones, and has been adopted by engineering founder societies such as ASCE. Short as it is, the Guidelines to Practice or Guidelines for Use each society feels compelled to adopt, in order to enforce it, lengthens it considerably (Appendix B). The eleven principles and canons compare favorably with the Ten Commandments Moses cited or the two Christ used. The difficulty begins when one tries to interpret, to apply or to enforce any one precept to a specific case.

Codes could be typified as exemplary or enforceable, as professional or business, or as voluntary or mandatory.

---

[b]Partial summary of sections of Reference 139.

Acknowledged professions which choose to adopt and enforce codes of conduct should educate and then admit only those willing to abide by their ethical standards. But even if any group could hope to enlist only such perfect professionals, their practice could at times be interpreted differently and thus require investigation. Provision would also have to be made to exclude from practice those professionals who do not qualify, even if other organizations or industries may address them as such. The code would be of no real value in protecting the public health and safety, as well as the public purse, if all who wished were allowed to practice. Hopefully, most members will choose to obey. Policing any code--or law--becomes impossible when a majority, or even a large minority choose to "disobey." All of civilized society is founded on the promises men are willing to make and keep, and on the enforcement of those covenants.

The way codes are typified and interpreted, or even prescribed in their Guidelines to Practice, depends to some extent upon the engineering specialty involved. Kemper has noted that "mechanical engineers, in discussing professional responsibility, often appear to emphasize product liability. Civil engineers are deeply concerned about personal integrity, and the elimination of conflicts of interest. Electrical engineers seem to spend much of their concern on corporate responsibility, and upon strengthening the position of the engineer vis-a-vis the employer..." (140).

Several engineering organizations have adopted creeds to guide the ethical behavior of their members. These are exemplary codes accepted voluntarily or merely noted as being desirable. For instance, the *Ritual of the Calling of an Engineer* founded in Canada in 1926 and the *Order of the Engineer* founded in the United States in 1970, do endeavor to "instill a consciousness of belonging to one another, to themselves as individuals, and to those whom they serve." These organizations require only a dignified ceremony at which the engineer pledges to abide by their Obligation (Appendix C) and to wear an iron ring on the small finger of his working hand. No action is taken against those who may violate this obligation later.

The *Engineer's Creed* (Appendix D) adopted by NSPE and *The Faith of the Engineer* (Appendix E) adopted by ABET express similar ideals. One of NSPE's Professional Policies "encourages the scheduling of public ceremonies at which its members witness new graduates or engineering society initiates accept such creeds voluntarily." However, acceptance of the Creed is not required for membership and so is merely accepted as a desirable ideal by most engineers. The Creed was written in the early 1950's by Paul Robbins, NSPE Executive Director, 1946-78, and adopted by that society in 1954.

These obligations and creeds cited above resemble the Hippocratic Oath (Article 7B-Paragraph 1). They are identical in intent but differ in their acceptance. The Hippocratic Oath was, until recently at least, required of all graduating medical students, whereas the engineers allegiance was entirely voluntary.

7E. ENFORCEMENT MECHANISMS[c]

Whether the objective of ethical code enforcement should be punishment of the violators or correction of a wrong depends upon the organization's philosophy and the criteria specified in its bylaws. Punishment is the more usual mode and may involve expulsion, suspension or a reprimand, sometimes with simultaneous public announcement in the organization's publication. Punishment probably deters undesirable behavior enough to discourage violations. Unfortunately, punishment for ethical code violations has been invoked quite sparingly as will be seen later. In more than one case, some societies have refused "to get involved" when their members appealed for help (54).

ENGINEERING SOCIETY ENFORCEMENT

Some professional societies, like those of economists, have no ethical codes. They believe that civil law should be used to punish those who violate legal constraints and that the sole purpose of corporations, in our free enterprise system, should be restricted to that of maximizing profits within the legal constraints imposed by society.

The point here is, who blows the whistle? Who is to determine when one of his fellow professionals, or of his corporate superiors or public servants, pursues programs that are not in the public interest or even unlawful? These actions could involve bribery or ex-

---

[c]Partial summary of sections of Reference 139.

tortion connected with public construction or procurement, or accepting substantial financial favors as a purchasing agent, or producing unsafe consumer products, or plagiarism, or making slanderous statements, or attempting to supplant other firms or individuals already operating under contract, etc. If professionals fail to report violations, they fail to accept that obligation expected of them by the public and perhaps of their ethical code as well.

Frequently, these violations of laws or ethical codes are discovered by our free press. Sometimes charges are brought by one or more members against their professional peers. Perhaps only a minor fraction of all violations are uncovered. Nevertheless, all charges must be investigated by those professional societies whose members are involved, if indeed they do have codes of ethics. Not all societies do.

Such investigations can be time consuming and lengthy. They should be thorough for men's careers may be at stake. They may involve delays caused by court cases on which evidence is based. They will involve considerable expense for ethics committee travel, society staff support, and subsequent per diem and travel for board of direction hearings. Legal counsel will usually be involved. Budgets for such expenses, including legal fees, exceeded $50,000 in 1974-75 for the Committee on Professional Conduct (CPC) of the American Society of Civil Engineers (ASCE).

The number of cases involving alleged ethical code violations of ASCE members for the period 1959-1974, and of the disposition of these cases was reported in its Manual on Professional Civic Involvement (142), and is set forth in Table 7B. The post-Watergate influence is evident. About 60 percent of cases were dropped for lack of evidence. Those members finally charged with alleged violations do have an opportunity to defend themselves before the ASCE Board of Direction. They may be represented by counsel. The Board frequently publishes pertinent details of the charges and punishment of those found guilty in its monthly magazine CIVIL ENGINEERING. In 1978 the ASCE produced an hourlong videotape depicting a mock ethics hearing entitled ETHICS ON TRIAL. It illustrates how the Committee on Professional Conduct presents cases before the ASCE Board of Direction (144).

TABLE 7B - ASCE PROFESSIONAL CONDUCT CASES
1939-74: (139)

CASES	1939-50*	1951-62*	1963-74*	1973	1974
Received	41	88	161	15	64
Discussed	61	138	240	20	82
Dropped	28	37	108	3	40
Hearings	0	19	28	3	12
MEMBERS					
Admonished	7	14	14	0	2
Suspended	0	20	9	3	7
Expelled	4	7	7	0	5
Resigned	2	1	3	0	0
Pending	21	56	100	15	31

*NOTE: Data represents the total number of cases for the 11 year period.

ASCE members against whom charges are filed for an alleged violation, or who have been advised that CPC is investigating the possibility of their having violated provisions of its Constitution, Bylaws, or Code of Ethics may not resign until the case is closed.

Despite comments one may hear to the contrary about the miniscule value of professional society membership, my experience indicates that virtually all members so charged fight to defend themselves and to retain their membership.

The National Society of Professional Engineers (NSPE) operates somewhat differently. Its member state societies conduct the investigations and take final action. When complaints involve interstate cases that are lodged directly to the National Society, the case is first referred to the appropriate State Society for a recommendation. The National Board then reviews the evidence and the State Society's recommendation. It then makes its own confidential recommendation to the State Society, but may elect to publish the results of its action in its magazine THE PROFESSIONAL ENGINEER. The final action by the State Society may or may not coincide with that of the National Board. Complaints involving only members at large, who do not belong to any State Society, are handled entirely at the national level. There have been only two hearings between 1947 and 1978 before the National Board. The NSPE Board of Ethical Review renders opinions on cases involving a

178

State Society's own members if invited to do so. It publishes many of these cases, including the facts but deleting names, with its recommendations, as examples in the Society's magazine.

It is difficult to determine how many NSPE cases are handled on only a state basis. Frequently they are handled informally and the members may be allowed to resign before formal charges are filed. It has been estimated that 150 cases per year are considered. Of the 55 member societies, 20 have never processed any disciplinary case and 15 are very aggressive. For instance, in 1974, the Florida Society investigated 45 cases and Iowa about 30.

STATE REGISTRATION BOARD ACTIONS

The discussion so far has indicated that engineering societies could only expel, suspend, or reprimand members who violated their codes. They cannot prevent a member--or non-member--from practicing. They could censure corporations or governmental bureaus for conduct not in the public interest or they could refuse membership to employees of such organizations. These societies never have done so. Perhaps they never will. But, it is possible that the future may see some of the societies joining in class action suits restricted to unsafe products or public projects. Hopefully, such legal entanglements can be avoided by prudent ombudsman counsel of technological leadership properly advising corporations and governmental bureaus of the engineer's responsibility to the public. Whether or not the engineer is excluded from licensure because of industrial and governmental exemptions may not apply in such suits. At any rate it may be far cheaper to adopt policies that involve corporate social responsibility than pay legal fees involved in the class action quagmire.

But what of those engineers who must be registered to practice? Their licenses always could be revoked by the registration board in all states for incompetent practice and for gross misconduct--but misconduct was not defined in the laws until recently, (Article 7B--Paragraphs 5 and 6). Some states have amended their laws and these boards have adopted rules for suspension for ethical as well as for legal violations.

The National Council of Engineering Examiners is unable to estimate the number of cases investigated or

179

of licenses revoked by the 55 State Boards because of legal prohibitions on the release of such information.

7F. CODES VS. LAWS VS. REGULATIONS[c]

Ethical codes define an acceptable set of behavioral standards that the public expects a group of experts, upon whom it has conferred professional status, to adopt and enforce. The major emphasis of such codes has changed since about 1950 to defining the engineer's paramount responsibility as being the protection of the public's health, safety and welfare.

Likewise, all licensing laws are predicated on the assumption that the restrictions they impose upon the practice of the practitioners involved are necessary to protect the public health and safety. Laws and codes, thus, serve an identical purpose. Laws require evidence of competence for licensure. So do professional societies, but not all have ethical codes and not all enforce these effectively.

When either the licensing boards or the professional societies falter in their obligation to protect the public, and particularly when industry does so too, the public insists that government agencies regulate the practice of the art as well as the working conditions, products and services of the industries involved. Thus all become constrained.

If the public is to be protected by legal constraints or by professional devotion, the members of society and of that profession must be willing at times to "blow the whistle" (53,54) on those few offenders who choose not to conform but who did manage to survive its rigorous admission standards. Initiating such charges requires very mature judgements. Some way must be found whereby concerned, ethical, and even idealistic engineers can avoid the quagmire encountered when they believe there is an ethical violation, but are unable to recognize the effect of conditions which ethical, well-established, experienced engineers accept as justification for a different conclusion. Employers and agencies have a need to communicate better with their professional employees in order for all to realize the assumptions and risks underlying a project.

---

[c]Partial summary of sections on Reference 139.

Every engineer must avoid all risks that may cause loss of life. But there is some risk in every design. Safety must be balanced against economic feasibility, as Binger observes, for absolutely stable road construction in mountainous regions of South America might be impossible to finance and would thus never serve the inhabitants (143).

Engineers of the future will at times need to adopt an adversary role to protect the public health and safety, and its purse and resources, if this civilization is to survive. Admittedly, technology creates ecological and social problems. Only technology is able to solve these problems. Engineers can help effectively only if they become activists and leaders in society. If they prefer the status quo, or if, as Santayana said, they cannot remember the past, they will be condemned to repeat it.

8

Engineering Management and Societal Leadership

8A. MANAGEMENT OF IDEAS VS. PEOPLE[a]

If the obligations implied in the public purpose of the learned professions are to embrace training for leadership roles as well as for industrial management and professional practice, new educational policies must be implemented. Except for the service academies of our armed forces, practically no collegiate programs now include courses in societal leadership as distinct from personnel and corporate management. Courses in engineering management and in business management are overly plentiful, but few engineering students study them. The principles of leadership and of management can be taught, but that does not mean that all who study these will become leaders or managers any more than do all who study mathematics become mathematicians. Leadership requires the ability to manage ideas as well as people, to apply one's professional practice within legal and ethical bounds, and to have the wisdom to select appropriate goals. Table 8A shows these needs.

TABLE 8A - PROFESSIONAL OBLIGATIONS (145)

I. To Manage Ideas
   a) Requires broad formal education for skillful practice.
   b) Requires lifelong education for continued competence.

II. To Manage People
   a) Requires cooperation with superiors, peers and subordinates to fulfill assigned missions.

---

[a]Partial summary of sections of Reference 145.

b) Requires persuasion of public bodies to formulate appropriate technical and/or societal missions.

III. To practice the Profession's Art within Ethical and Legal Bounds
  a) Requires a sense of mission.
  b) Requires professional freedom to act without unnecessary constraint.

Many professionals cannot lead because they were never groomed for the role, or will not because of the bureaucratic constraints now prevalent in government, as noted by Warren Bennis in THE UNCONSCIOUS CONSPIRACY: WHY LEADERS CAN'T LEAD (146). The leadership role will involve service as the public's advocate or lobbyists adversary on professional questions affecting society. The role need not necessarily require service of the professional as a legislator. It will require a united profession whose advice on professional matters will at least be considered by lawmakers. Essentially the innovative engineer is an artist. Is he to paint the pictures the museum wants, or those his talents can master?

Ethics as well as laws are needed for such professional practice and leadership. To be effective, both ethics and laws must be observed. If they are, there is no need for enforcement. Codes and laws can only set limits beyond which behavior is condemned. Barzum summed this up for professionals by indicating that "Policing is not enough...moral regeneration can come...only when...members feel...confident that behavior is desirable" (21). Almost all professionals do practice legally and behave ethically, but many are unwilling to discipline those few of their peers who violate accepted and prescribed limitations. All engineers need to see laws and codes enforced, but also need to know that they can expect the support of their own professional societies when questionable cases arise, or when their technical judgement is overruled by non-technical management and public safety is endangered.

Professional practice alone may span the spectrum from purely technical functions through the management of engineering projects to the establishment of corporate or societal missions. Many of the personal attributes needed to fulfill any of these three roles are identical. Characteristics like integrity, persever-

184

ance, optimism, responsibility, dedication, and a willingness to act come to mind. Others will be discussed later. Note, however, that those which relate to leadership, as distinct from personnel management, involve, in addition, wisdom, optimism, compassion, inspiration and a sense of the social importance of proposed goals.

Edward Wenk suggests in MARGINS FOR SURVIVAL (147) that we professionals must team up with legislators to establish national goals and ask "Can we do it? Ought we do it? What happens if we do it? What happens unless we do it?" He stresses the need for professionals to anticipate and to accommodate the uncertainties of their practice.

Far more planning will be needed in the future by professionals if our living standards are not to retrogress. However, good management will delegate such authority and responsibility as far down the line as possible. Nevertheless, planners will have to think globally as they practice locally if they are to avoid the creation of super states and strangling bureaucracies.

"Small is beautiful" might be applied to governments to provide more efficient service. Then the professional standards could be upheld by professional ethics and peer pressure instead of the ill-fitting legal constraints imposed by the regulatory agencies of overgrown governments.

All learned professions of the future will have to broaden their public purpose to include societal leadership, for only they can supply the knowledge needed to change or establish national technological policies. All professionals will have to embrace team work for all societal goals involving complex interrelationships. No one "pro" knows it all.

Frederick Mavis expressed these thoughts so conclusively for engineers; i.e., professionals, when he wrote, "Has not the time come when professionals may have to study the laws of politics--and harness the overgrown giant of governing machinery which has become America's master, rather than its servant? Would professionals bungle the job more than politicians have done?...The job of harnessing politics for the use and convenience of man has yet to be done?" (148).

Changes in society determine the type of leadership needed to operate it. The demand for political

185

leaders rose sharply in the last half of the 1700's. They emerged and produced a few governments which guaranteed freedom to speak, to produce and to trade. That political environment nurtured the newborn Industrial Revolution and accelerated rapid, worldwide technological changes. Now, there is an urgent need for a new type of leader who, as Brown indicated, understands science and engineering (11), and who as Wenk has suggested, is also aware of the social responsibility that a professional should assume (147).

Alvin Toffler, author of FUTURE SHOCK and THE THIRD WAVE said, "This is not the first time we humans have been hit by a giant wave of change. It first happened after the invention of agriculture 10,000 years ago, and that led to the first civilizations. It happened again 300 years ago as the industrial revolution spread a better way of life...(We) can prepare for such changes...Changes becomes less confusing as we see how isolated changes relate to each other and merge" (32).

If engineering professionals can be impressed with the need to participate actively--not passively--as leaders in the formulation of appropriate missions for our technological civilization, it will have a better chance of progressing more rapidly and equitably. Such activity need not mean that the engineering professional must run for public office, although it must be admitted that too few of his peers do.

If an engineer would lead in the public sector he must secure a power base; i.e., be elected, be appointed, develop a following (like Nader), head a professional organization (like ASCE, AAES, etc.), or testify as an expert witness on societal issues (156). However, few engineers will have much opportunity to achieve any of these five bases. Society needs desperately to base technological policy on engineering expertise if its free enterprise system and political democracy are to survive. The engineer's objective counsel and logic will recommend far more practical solutions for societal issues than those solutions based on emotion-like appeals.

How, then, is the engineer's voice to be heard? He will have ample opportunity to exercise individual leadership on local issues. On national and global issues he will be able to exercise collective leadership only through some unity organization, but then only if he has a direct voice in the formulation of positions on public issues. If these issues are only discussed in committees until tired hands are raised above tired heads, little will be accomplished.

The several engineering technical societies represent a mode by which members might have a voice on national and international socio-political-technological issues if their elected officers operate from any of the five bases listed previously. Even then, the action of these officers would generally be restricted to the weakest mode; i.e., that of appearing as an expert witness before legislative committees. But the action then would be muted by the fragmentation of the many engineering societies. Such testimony would be more effective if these societies were united, not only in a federation as AAES is now, but also as an umbrella organization of individual members who have the right to participate in the formulation of its policies on major socio-technical problems.

8B. ENGINEERING VS. BUSINESS MANAGEMENT VS. GOVERNMENTAL PLANNING[b]

Engineering management and business management are based on many of the same principles, yet they have different objectives. Engineers manage technical projects whose purpose is to design and produce products, and to build and operate systems. The job of securing the required materials and manpower needed to produce goods and services, as well as to distribute and sell them is the fort of the business manager in free market economies. In socialistic governments such planning is relegated to bureaucratic control, since the governments own the means of production. The engineer tries to fulfill his assigned technical mission by minimizing materials and maximizing production. The corporate executive tries to fulfill his role by minimizing labor and maximizing profit. Governmental bureaucracies are handicapped in that objective because of the political pressure to maximize employment.

Certain types of materials and energy like iron and oil tend to become scarcer as their industrial use escalates. They will always be available at a price. Barring innovative mining and drilling, the costs may escalate to a point where the end use of these materi-

---

[b]Partial summary of sections of Reference 90.

als change or are replaced by substitutes. For instance, if oil became scarce enough and it took more energy to produce a barrel of it than it contained, it might be used only as the raw material for plastics and medicines rather than fuel. Energy then might have to be obtained from solar collection or nuclear fusion. Iron, likewise, might be replaced by composites.

Engineers have shown a far greater capacity and/or willingness to manage technical projects than to lead in establishing public missions. Florman (4) wonders if engineers are too unorganized to lead effectively... and too honest and naive to function competitively. He pleads for broadly educated engineers to assume more of a leadership role in the "establishment" as they cope with the politicians and entrepreneurs who run the world.

One might argue that although engineers have not fulfilled their professional obligation in societal leadership roles, they have succeeded superbly well in their technical role. Our system has developed "isolated, topflight, scientific and engineering programs whose graduates perfected radar and miniaturization in the 1940's, transistors and nuclear power in the 1950's, calculators and computer graphics in the 1960's, and micro-processors and structural composites in the 1970's. These graduates even sent men to the moon and back!" (155). They represent a small elite--a dedicated meritocracy--whose successes were accomplished in spite of the inertia of the masses of their fellow men, and even of some segments of their peer groups (109).

We still need such a technically competent, dedicated elite to recognize the priorities in our technological civilization, and then to inspire and manage the diverse popular and political support required to implement long term solutions. Wenk (147), like Florman (4), pleads for more sensitive technological leadership...

Despite the phenomenal achievements since 1940 in the technical sector, in the political sector we find an increase since then of governmental regulatory agencies which constrain production and reduce our ability to compete in world markets. Some of this interference is related to the difficulty of developing a consensus on technological issues (Article 5E, p. 105), particularly when they are controversial. During this same

188

interval, engineers have made little effort to engage in societal leadership, or to educate their novices in it and in the principles of engineering management as well.

Tribus (156) believes that engineers constitute a self-selected body and are taught to serve as team players--to be on tap. Conversely, lawyers are taught to be adversaries--to be on top. He also feels that engineers are best able to see into the future because they rely on natural laws, can measure almost everything, and can make crude models of even social behavior. The curricula which prepare them, however, stress logical analysis and instill impatience with ambiguities. Curricula emphasize how to solve rather than how to formulate problems. In the societal sector ambiguity is all too prevalent. Thus the engineer is at a disadvantage as he tries to lead. He would rather join than fight; to tell all in legal cases rather than only what must be known.

Perhaps, as Goshen (157) finds, the traits many engineers possess, like precision, perfectionism, humility and conservatism, as well as an inclination to manage ideas rather than people, and to expect people to obey laws as precisely as objects do, militates against the engineer's ability to lead. Perhaps so? However, all students can be made aware of their professional obligations to contribute what leadership their talents allow.

It does little good to possess leadership potential and to wait--like a bashful maiden--to be called upon to use it. Nor is it productive to assume a position of leadership and then not to lead. The first represents an opportunity only. The second implies action and the likelihood of encountering opposition. An intimate understanding of technology is vital to direct intelligent management of corporate enterprises; it is also needed indirectly for statesmanlike leadership in the political arena.

How, then, can engineers assume positions of leadership? Success in engineering usually "leads to management positions whether the engineers aspires to them or not" (157). The principles of societal leadership differ somewhat from those of engineering management. Table 8A indicated that the critical difference between the two types of leadership centers on item II-b; i.e., on the ability and willingness to formulate appropriate societal, as well as technical missions.

The technical achievements of engineers have been superb, but sometimes undesirable side-effects occurred. Engineers have not always been blameless of such developments, and cannot always fault business executives or politicians for unwanted developments. Industrial decisions are usually made by groups of experts in the technostructure. The economist John Kenneth Galbraith believes that top managers are not decision makers but decision ratifiers (41). Industry needs a group technostructure of engineering and corporate managers whose advice on technological issues will be also be ratified by (politicians) statesmen.

Engineers, according to Dallaire (41), constituted 14% of management in 1900 but, by 1964, this had grown to 36%. Actually the percentages were higher (53%) in younger men. More recent surveys by Heidrick and Struggles (154) show that in 1978, 68% of the 500 largest companies had technical representatives on their boards of directors at the corporate level. Less than 2% of R&D directors had no college education, and about 90% were either engineers or scientists, 56% of whom had doctorates. The 1981 educational levels of all board members are listed in Table 8B. About 55% had post-baccalaureate training. Similarly, for the 1000 largest companies, only 9% of their chief executive officers (CEO) in 1980 had no degree, 46.9% had a B.S. degree and 43.8% an MBA. The proportion of MBA's had increased from 1975 when it was only 37.6%.

TABLE 8B - EDUCATIONAL LEVELS OF BOARD MEMBERS (154)

Type	No Degree	BS	MBA	J.D.	MS/Prof.	Ph.D.	Total
New	8.6	35.5	17.5	14.2	8.7	15.5	100%
All	10.0	35.1	17.7	11.8	9.2	16.2	100%

The survey also indicated that R&D executives who considered new product development important had increased from 40.7% in 1973 to 67.7% in 1978. Similarly, those who considered delaying product obsolescence less important decreased from 10.2% in 1973 to 1.9% in 1978. By 1981, 76.2% agreed with Japanese criticism of American business that there was an over-emphasis on short-term profits, and 61.4% believed that managerial control of quality was lacking.

Corporate objectives in the United States began shifting in the 1970's from investing in plant modernization so as to compete more effectively in world trade, and in plant expansion so as to produce more goods, to corporate mergers so as to diversify sources of income, increase profits and reduce income taxes. One might not fault business managers for this change because the economic constraints and/or incentives formulated by Congress encourage the shift, at least for the short term. Nevertheless, the change created no new jobs; in fact it exported jobs as foreign competition moved quickly to adopt modern technology and to produce superior products.

Reginald H. Jones, who ran General Electric Company for nine years until his retirement in April 1981, spoke for business in Washington as a leader of its primary lobbying groups, the Business Roundtable. His departing recommendations were, however, directed more toward business, which he felt had opted for short-term profits which obscured the necessity of building for the future. "While we were resting on our laurels, companies abroad were taking the long-term risks," he said (59).

William C. Norris, Chairman of Control Data Corp., told a House subcommittee in 1979 that, "Most large companies create a corporate bureaucracy that avoids risks. The emphasis today is on immediate payoffs (and)...development of new products and services takes a back seat. Similarly William J. Abernathy, a professor at the Harvard Business School said, "The figures need to be pointed inward. We got...lazy and went to sleep" in the 1960's and 1970's. "By their preference for serving existing markets rather than creating new ones...(American business managers) have effectively forsworn long-term technological superiority as a competitive weapon" (59).

U.S. productivity has fallen sharply since 1973 and, in the 1980s, lagged far behind levels in Europe and Japan. Research, which grew rapidly through the 1950's and 1960's, has declined and also moved from basic to applied research and development. Abernathy and his Harvard colleague Robert Hayes also felt that "Mergers...tend to produce quick and decisive results, and offer the kind of public recognition that helps careers along" (59).

Mergers in the U.S. rose from $12 billion in 1975 in about $7 billion yearly increments to $40 billion in 1979. They produced few new jobs. In 1982, U.S. Steel Company acquired Marathon Oil Co. in a merger. Had it spent half the amount needed for that purchase, it could have completely modernized all of its mills. According to Texas Senator Lloyd Bentsen, part of the problem resides not in public policy, but in corporate boardrooms. Japanese firms secure far more of corporate financing from debt rather than equity, pay bonuses based on returns over a period of years, retain the service of their top managers longer, and generate greater employee loyalty than American firms do (61).

Perry Pascarella (158), Executive Editor of Industry Week, believes that "Profit and productivity are only symptoms. What's really needed is new top management attention to people and products. And that calls for a new type of executive...American (industry) is coming to suspect...some flaw in U.S. management. Some of it has been directed at the...MBA programs...accusing them of too much short term, financial orientation." He feels that some firms (like GE), reaped the benefits of technological renaissance because they paid attention to R&D and were entrepreneurs. "In the 1950's, industry's top management was generally production oriented. By the 1960's the companies that were growing the fastest were...market oriented."

There can be little doubt that the primary need for any industry in any economic system is to operate at a profit. Even communist nations which borrow money to expand their industry are expected to repay it. McKetta (78) feels that "The free market can be a socially brutal system (but)...so is slavery...So is excessive governmental regulation—just look at Russia today—or even England." He says it now takes the U.S. 13 years to build a nuclear power plant, whereas Japan and Europe do so in 4.5 years. Our first one took us only 4 years! The first one (British) at Calder Hall took only 1.5 years, but its capacity was only 50 megawatts.

Whether industrial managers in the free world have an easier time of it than those in communistic regimes is questionable. At least ours realize how complex the free market is, do not try to ignore its irresistible force, and try to work with government in establishing pollution and safety controls, as restraining as these can be. They are also more successful than are those

192

in planned social orders, as a 1982 World Press Review (159) note indicated. It reported that

"In Poland's Katowice region near the Czechoslovak border...the sulphuric and nitric acids that fall with each rain have so corroded railway tracks that trains are restricted to 25 miles an hour...The Karowice region covers only 2% of Poland's land area but contains 10% of its population. The people of the region have 15% more circulatory system disease, 30% more tumors, and 47% more respiratory disease than other Poles... Every urban area...gets more than the permitted annual dust fallout of 250 tons per sq.m...The lead content in soil samples is hundreds of times the national limit,...and cadmium 4 to 16 times the prescribed limit, according to the report (of the Polish Ecological Club) which recommended that at least 17% of the region's farmland be taken out of production...Katowice may have the worst pollution problem in the world according to (the head of) the Institute of Environment Engineering of the Polish Academy of Sciences. (Local) officials agreed in theory that they have the power to close factories and have listed 44, but (they plead that) 'It is not correct for regional officials to take such actions'...Farmers have given up growing sugar beets because the leaves poisoned livestock...Not surprisingly, (workers) are the worst affected. In 1971-80 only 21%... left...because of normal retirement; more than 80% were given disability pensions (in 1980)" (159).

One can only conclude that our free market constraints and governmental regulations are not so bad after all. The point is, however, that we could manage our economic system even much better than we do through greater voluntary cooperation of all concerned. It seems obvious to me that economic nationalism is incapable of providing adequate public safety or consumer goods or durable peace. Of course all economic systems for the production and distribution of goods and services are, as Thomas Sowell points out in KNOWLEDGE AND DECISIONS (160), designed to ration inadequate supply. All systems have procedures for preventing some people from getting what they want or need. All managers try to match needs with wants, but human wants are insatiable. All governments jail thieves; all ration supply, providing money, coupons, or food stamps as mediums of exchange.

With all of its faults, the free enterprise system is the best managed so far. Its improvement presents an everlasting challenge. Its profits and losses play two key roles; they provide incentives to produce, and they direct resources into the hands of the most competent (161). Resources shift out of over regulated and price controlled sectors of a nation's economy into less urgent uses like subsidized industries. Sad to say, a dollar spent lobbying brings greater return than a dollar spent in production (162).

8C. PRINCIPLES OF ENGINEERING MANAGEMENT

The discussion so far dealt with management of ideas vs. people, and on the differences between the objectives of engineering and business management vs. government planning. A precise definition of engineering management and a general discussion should, perhaps, precede a listing of its suitable principles although most of these will be alluded to in the illustrations. Merritt Williamson's definition, taken from the ENCYCLOPEDIA OF MANAGEMENT, is complete and concise.

"Engineering management is...concerned with managing technical work and technical people in a predominantly technical environment...(It) is the art and science of planning, organizing, allocating resources, directing and controlling activities which have a technical component. It differs from Industrial Engineering...by its greater focus on people problems rather than on system design...It differs from general management in its requirement that practitioners be competent in some technical field" (164).

A 1969 Engineering Manpower Commission survey showed that 80% of all engineers were *regularly assigned* managerial duties. These data differ from those in the 1968 NER and 1981 NCEE surveys shown in Table 6A which list 35% *engaged primarily* in management. The discrepancy is due, probably, to the amount of time devoted to management rather than to design, operation, construction, etc.

194

Williamson differentiated between engineering vs. administrative management by noting that the former "is really the management of important and scarce resources, including people of (technical) ability...with a great deal of concern for the well being of society... Engineering management is (an) art and a science." The science provides tools like operations research, statistics and probability,...programming,...general systems theory, etc. Administrative management involves the planning of the system of accounts, interpretation of cost data, budgeting, scheduling, approving expenditures, progress reports, authorizing overtime, etc. Williamson also feels that all specialists need a manager to make his work more effective (165).

The difference between engineering management and other kinds was also described by Dougherty (166). He compared it with management in industry and government. Both creep along the paths of bureaucracy as they age.

Both "believe that any competent manager can manage anything...(in) government, efficient operation is not a first requirement... In industry, cost-effective operation is a requirement for survival...(present) trends of elevating financial and legal personnel to (be a) chief executive officer (CEO) reflects the belief that training and experience in... production processes are unnecessary...The normal evolution of established firms is 1) to maximize short term profits as soon as (the firms become) established, 2) to cut prices to retain a share of the market (or to establish a monopoly), 3) to (cut R&D expenditures) and 4) to ignore long-term goals.

Modern industrial corporations are basically technical in nature...(were founded) by knowledgeable entrepreneurs and managed by technically knowledgeable managers... (who) mastered the art of successful delegation to responsive and capable people...Engineering management...(is) the management of technical enterprises and technical personnel...(This) cannot be done over the long run without knowledge and understanding of things technical. (It requires turning) technical people into managers...the reverse is much more difficult...Unfortunately our very best engineers make very poor managers...

195

The industries not doing well seem to have forgotten the (paths they followed) to become giants...U.S. Steel managed itself steady-state for decades, and General Motors was not far behind...The chief executive officer (should spend) the bulk of his attention on the future...to visualize both the probable and the possible...Extrapolation of the past will not permit changes" (166).

Dougherty compared steady-state vs. long-range management for autos, planes, steel and computers. He noted that, in constant 1967 dollars, the cost of cars increased 1/3 from 1950 to 1980 due mainly to a real increase in performance, whereas the 1980 F-15 cost 70 times more than a 1942 P-51, but it could clear the skies of all P-51's. Over a 70 year span, air transport shifted from open biplanes to the space shuttle. Similarly, the cost of steel increased from 1950 to 1980 by 76% due mainly to its increase in quality, while the cost per bit used for computers decreased from 17.8¢ in 1955 to 0.004¢ in 1979 (166).

The foregoing comments illustrate the difficulty of managing technical enterprises for long periods using too small a proportion of technical professionals. The challenge is to find, or train, engineers or scientists in the art of management. One practical solution would be to teach the bare fundamental principles of engineering management to all engineering students just as they are taught the essentials of the engineering sciences and design. Both disciplines could be supplemented by continuing education later as the individual's talents and desires dictated.

The principles of engineering management must be based upon a thorough introduction to basic science and engineering as well as to operations research, systems theory, etc. to manage ideas. The principles can be applied only by convincing and/or motivating one's superiors, peers or subordinates. A few of the means for accomplishing that objective follow; they reflect common sense, yet they are easily overlooked or forgotten.

1. Convey a sense of mission.
2. Keep your superiors aware of your plans.
3. Prepare a worst case analysis. Prepare for trouble.
4. Keep things simple. Observe early warnings.

196

5. Seek help quickly.
6. Take reasonable risks. Don't become obsessed with failure.
7. Cultivate how to cope.
8. Don't lose touch with your co-workers. Listen to their opinions.
9. When you must criticize, try a small dose first; help associates, don't humiliate them in front of their subordinates.
10. Motivate subordinates by assigning challenging opportunities. Praise their achievements.
11. Be fair--but strict--but fair.
12. Strive for excellence.
13. Make decisions. Accept responsibility.
14. Issue directions that permit acceptance without resentment.
15. Recruit or train capable subordinates to prevent group failures. Toscanini could not conduct superbly orchestrated music from a high school band.

The list is seemingly endless. Management's task is to organize a productive enterprise, direct people to fulfill assigned missions, utilize available talent, and provide opportunity for development. Engineering managers should convince subordinates that they would not ask them to perform assignments they cannot do themselves, or would not do if they had the time and ability; that they are doing everything possible to insure success of the assigned mission. The first responsibility of any group and its leader is to fulfill its technical mission. An auxiliary responsibility for professional leaders relates to the assistance they can render to formulate suitable corporate or societal missions.

The important challenge is how to utilize engineering personnel to upgrade the management structure. The Japanese have shown a talent for that transformation as well as for decentralized decision making. They delegate authority and responsibility effectively. Many Japanese workers feel responsible for quality control, cost reduction and productivity improvement. At Toyota, for instance, 9 suggestions per employee per year are received and 83% are implemented; at General Motors the average worker makes about 1 suggestion per year, but only 20% are adopted. The Japanese invest heavily in applied R&D in organization research (167).

197

8D. PRINCIPLES OF SOCIETAL LEADERSHIP

A familiar cliche reminds us persistently that leaders are born, not made. If so, astrology or heredity might help. However, individuals blessed with any natural talent will be better able to display it if they are also taught the principles involved in its application. A gifted pianist, for instance, with a distinctive flare for interpretation and touch, will be a better performer if he has studied harmony, counterpoint, and musical composition and practiced scales under the supervision of a teacher dedicated to the pursuit of excellence.

It would, therefore, seem obvious that university objectives would recognize their need to train all gifted students with a potential for leadership in leadership as well as in their professional specialty. Few schools have any leadership courses or programs, yet all colleges recognize their need to transmit existing knowledge, to search for new knowledge, and to motivate students for continued, life-long study. How successful this educational effort has been depends upon how optimistically or pessimistically one views world events. Currently (1983), events appear to me to be both politically disastrous as well as technically fortuitous. These were caused by societal missions formulated by "educated" people in governments, industries, churches, and schools. There definitely is room for improvement, but then more wisdom would have to be involved. The confusion was best expressed by T. S. Eliot as:

"Where is the wisdom lost in knowledge?
Where is the knowledge lost in information?"

to which Edward Cornish, president of the World Future Society in 1981, added the more modern version of:

"Where is the information lost in data?
Where is the data lost in computers?"

Wisdom is, thus, a most desirable attribute of all who aspire to leadership. Wisdom is preceded by knowledge as well as by experience. All three permeate every principle of leadership, a few of which will be listed later. But so should moral and ethical principles be involved in the formulation of societal missions. Nelson and Peterson argue this thesis convincingly in a series of articles for engineering practi-

198

tioners in which they stress their need to act as moral agents; to recognize the conflicting allegiances they owe to their client or employer, their peers, their families, and the public; to distinguish between an explanation of why something occurred and a justification of why it was right; and to realize the distinction between moral, immoral and nonmoral judgements even in cost-benefit analysis (149).

A similar study of the social sciences also reveals that mankind appears to be adopting a new ethic and a new value system. Jonas Salk points this out by suggesting that society appears to be shifting from an epoch A to epoch B; i.e.; from competition to cooperation. In A, elite groups were concerned with themselves; man tried to make life possible and then enjoyable. In B, elite groups are concerned with whole species and of man's relationship to man and to his environment. Youth particularly "is impatient with the slowness of change toward improving the quality of life, especially for the still disadvantaged" (150).

Similarly, a study of psychiatry, as Judd Marmor, Director of Los Angeles Sinai Medical Center indicated, would reveal that our effort to strengthen "the moral fiber of our youth, restoring the family influence, or reinforcing religious teachings...miss the mark because the problems threatening our survival lie not in individual psychopathologies but in our socially sanctioned...group values. It is not the 'defectives' among us but we, the normal ones...who fight wars, cutdown forests, pollute rivers,...discriminate against minorities, and pursue profits..." (151).

COMMUNICATION

On what, then, should the principles of societal leadership be based as the foundations of civilization appear to shift? Perhaps foremost is communication. Not only must a leader utilize all available channels such as TV, radio, the press and personal appearances. He will be a better leader if he is eloquent, and states his goals in terms simple enough to be remembered and precise enough not to be misunderstood. Winston Churchill spoke of "blood, sweat and tears;" he united an empire. Martin Luther King "had a dream;" he extolled civil rights. Not all leaders need to be orators of their rank, but serious study of the art of speaking is essential.

199

FLEXIBLE OBJECTIVES

A second important principle relates to objectives, to what is said as well as how it is said. Objectives should be stated concisely enough to be remembered, but flexible enough to fit changing circumstances. They need not be so vague that they resemble empty political promises uttered only to insure a candidate's election. But they should not be stated so completely as to leave no loose ends for compromise. Leaders must know when to stop talking, to listen, and to wait. L. Pearce Williams described this principle of flexibility in his very impressive paper on PARALLEL LIVES (152). He illustrated leadership in religion, science and politics not by listing its principles nor the attributes leaders possess, but by comparing three pairs of men: Desiderius Erasmus (1469-1536) with Martin Luther (1483-1546) who competed for reform of the Catholic Church; Antoine Laurent Lavoisier (1743-1794) with Jeane-Baptiste de Lamarck (1744-1829) who worked to create new sciences; and Maximilien Robespierre (1758-1794) with George Washington (1732-1799) who helped create new countries.

"Erasmus of Rotterdam...devoted his life to the Renaissance ideal of recapturing (the) antiquity...of the early Church...to return men to the worship of God...His book IN PRAISE OF FOLLY scandalized the orthodox with a scathing attack on the corruption of the clergy...Erasmus did not lead the movement he did so much to prepare...He...felt that reform should be left to men of action...It was not that he was a coward, but that he lacked the self-assurance and self-righteousness necessary to defy the established institutions of his day...

Martin Luther...was not half the scholar that Erasmus was. He was...a volcano of emotions...he saw that his salvation depended on God. This conviction...was the basis of Luther's 95 theses nailed to the chapel door at Wittenberg in 1517...He was ordered to recant,...(defied) the greatest earthly power,...and left Worms to lead the Reformation...It was the power of his word that broke the Catholic monopoly on salvation. Luther was a religious zealot; Erasmus a pious scholar.

Leadership in the sciences requires...a "fit" between theory and fact that no amount of rhetoric can provide...Choosing the right way (of convincing) requires vision and leadership...Erasmus Darwin in England and Friedrich Schelling in Germany had profferred evolutionary theories (that species were eternally changing)....at the close of the 18th century. Thus, when Lamarck offered his arguments for the transformation of species, the way--still blocked by Church resistance--had been prepared for him...What Lamarck did (in the early 1800's), and Darwin was sensible enough not to do, was to attempt to fashion a complete theory...Lamarck's chemistry of vital processes (electricity, magnetism, animal heat)...were the "substances" that organized organic tissues...He did not know when to stop...

Lavoisier (redefined) chemistry in the 1770's and 1780's...It had been a kind of molecular physics...(H)is theory of oxidation...banished phlogiston...Instead, he insisted that chemistry dealt with observable entities,...(classifying chemical elements) with no concern for explanation...He...theorized that combustion and calcination of metals were due to combination of combustibles and metals. The opposition assumed that (phlogiston was released)...He (founded) the Annales de chimie, and permitted chemists who made the switch to the new names to publish their papers in it...Lavoisier realized that once you convinced others to speak your language, they would have to use your theory...Although he was executed in the French Revolution, his disciples ensured that...teachers of chemistry in France had to follow the new chemistry...to be hired in state-run schools.

(Similarly), Sir Isaac Newton, as president of the Royal Society, used his prestige to favor candidates for teaching positions in physics who followed his new theories. In the modern world, Louis de Broglie, Perpetual Secretary of the French Academy of Sciences, frowned upon the teaching of modern quantum mechanics...

Lavoisier's success and Lamarck's failure underline two aspects of leadership. Lamarck made the mistake of forcing his readers to accept a complete theory with no loose ends...Lavoisier was careful not to claim too much, (and was shrewd enough) to make sure that the competition would have to play according to the new rules.

Leadership in politics is, (however), more complicated. Robespierre,...a provincial lawyer, showed no particular talent for politics or leadership...He was brought up on the ancient (Roman) classics, from which he drew his hatred of tyrants. He borrowed (Rousseau's)...notion that men were essentially good but easily corrupted...(that) the function of the state is to force men to be free...Seeing enemies on all sides, he eliminated 30,000 citizens...The Terror was (simple-minded) action and in times of crisis people want action. It stimulated loyalty... and patriotism. It unified France. Robespierre had a fatal flaw as a leader. He did not know when to stop...He lived in a world of theory (trying) to transform the French into...virtuous people...On July 27, 1794, Robespierre was dragged to the guillotine.

The contrast between George Washington and Robespierre was stark. Waahington was not an intellectual nor a theoritician. He knew how to inspire men...Honor, to him, was more important than life...he sought and took advice...did not feel that his was the only way,...was a man of indomitable will...he expected no miracles, nor did he push his countrymen" (152).

Luther, Lavoisier and Washington shared some common characteristics that made them successful leaders. They were courageous, dedicated, and gifted. These are necessary but not sufficient conditions. The same could be said about Erasmus, Lamarck and Robespierre. But they lacked the ability and willingness to sustain the loyalties of their followers. Leaders are chosen and followed because they inspire confidence, and not just because they present rational policies.

People know leadership when they experience it. Sometimes social conditions are such that they complacently follow the wrong leaders like Hitler or Stalin, or are forced to do so out of fear. Such "leaders" are practicing drivership rather than leadership. Their goals seldom endure. Genuine leaders who have compassion for their followers, as well as goals crucial to society's existence, start movements that continue long after they relinquish control. Christ, Mohammed, Buddha and Confucius were examples.

There is a difference between occupying a position of leadershp and being a leader. Medieval monarchs often were ill-fitted for their roles. But so are many contemporary managers who ascend the corporate ladder via the Peter Principle (153) until they reach their level of incompetence.

Society depends upon leaders to initiate desirable changes in social customs and creature comforts. These leaders represent a small, usually lonely elite. Their talents vary, but all possess an expertise not shared by laymen, and a determination to succeed. When their proposals for a change to a "better" way, or their ability to communicate with potential followers at their level of understanding, or their sincerity are convincing enough, changes do occur. General Bruce Clarke noted that "Men face the possibility of death in combat with high physical courage. But all too frequently, the moral courage to risk possible damage to their career for a rightful cause in peacetime is lacking in men of apparently great promise" (168). Leadership involves "commandship" as well as management.

The ability to chart new paths for men to follow in science, religion, economics or government, requires that men must, as Havelock Ellis noted, "turn one's back on men." Otherwise they would follow well-worn paths. If they lead effectively, they soon learn to concentrate on a single purpose. The late newspaper columnist, Walter Lippman, when commenting on the death of President Franklin Delano Roosevelt, noted that "The final test of a leader is that he leaves behind him in other men the conviction to carry on...The genius of a good leader is to leave behind him a situation which common sense, without the grace of genius, can deal with successfully."

8E. GOALS
Cecil Rhodes (1853-1902), the British entrepreneur
and founder of the scholarships which bear his name,
echoed the sentiments of many other leaders who dream
of changing the status quo. His last words were, "So
little done--so much to do." Setting goals is much
harder than wishing, but day dreaming has its purpose.
Implementing the goals is harder still. That takes
leadership or brute force.

Goals which become societal missions, be they
safety regulations or resource conservation or strip
mining, affect the whole economy. Frequently they conflict. For instance, the economist Julian Simon, in
THE ULTIMATE RESOURCE, explains that even though in the
short run resources are limited, in the long run human
creativity overcomes obstacles to economic growth
(169). "The main fuel to speed our progress," he says,
"is our stock of knowledge, and the brake is our lack
of imagination." I might add that hope is even more
essential. He disagrees with the gloom and doom pro-
jected by the Club of Rome forecast and described by
Meadows in LIMITS TO GROWTH (170) and by Vacca in THE
COMING DARK AGE (171).

Meadows' computer modeling predicted that dwindling resources and exploding populations would stop
all economic growth within a century; Vacca believed
that "our great technological systems of human organization and association were continuously outgrowing
ordered control (and) are now reaching critical dimen-
sions of instability." He foresees crises in communications, transportation, electric power, management,
with monastic communities trying to survive. Whether
Meadows and Vacca are pessimists or realists, I leave
to readers and time to resolve. Simon, however, notes
that food production in the U.S. is increasing even
though acreage is decreasing,* that natural resources
will constitute a smaller proportion of our expenses,
that pollution is decreasing, and that our standard of
living (in terms of life expectancy, material afflu-

---

*The U.S. Department of Agriculture reported that
between from 1974-78 cropland grew by 18.2 million
acres. Farmers were paid to idle 37-65 million acres
per year from 1950-1972 to reduce farm surpluses. By
1972, farm production was 167% of the 1950 level.

ence, and political freedom) has been improving even since the Industrial Revolution started. Extra children ultimately become producers and a few develop into innovators. He wonders at the distorted logic of people being horrified at starvation in distant lands while endorsing abortion; or of the logic of leaders praying for those fallen in battle who might have been another Mozart or Michelangelo or Einstein, instead of eliminating the causes of war. All of his observations could be, or have been, transformed into goals.

Similarly, Sowell (160) is concerned about socioeconomic issues that could involve technological goals. For instance, the use of crises to expand bureaucracies beyond the ability of citizens to monitor them, or of attempts of "social justice" to redistribute benefits in excess of their existence should be debated and resolved. Meadows (172) proposed three ways to change distribution patterns; 1) let the rich nations volunteer to give more to the poor nations, 2) force rich regions to transfer more to the poorer ones, or 3) redefine distribution—enjoy now, pay later, if at all. He favors creating a desire for society to live in as non-destructive a way as possible—to design products for permanence rather than obsolescence. If industry would accept that goal, society would save resources because things would last longer. I might add that although fewer man-hours would be required to make the goods, just as much would be available. That would require some modification in our economic distribution system. The standard of living would not diminish; people would have more leisure. But, can they be educated enough to enjoy it?

There are other technological goals which relate to legal rather than economic constraints. Whether these constraints relate to goals endorsed by the public, or to the over-activity of the surplus of lawyers noted in the later paragraphs, is open to conjecture. Big city slums, for instance, are not natural disasters like earthquakes. They are man-made, guaranteed to destroy the property tax base and to increase crime. American policy to tax property improvements and land values, instead of European and Australian policy to tax only land values ultimately reduces investment income to zero. Maintenance must then be deferred, buildings and neighborhoods deteriorate, and slums are generated. Foreign cities which avoid taxing property improvements have fewer slums, encourage new construction, avoid the need for rent control, and attract industry (173).

Another legal constraint relating to production involves the minimum wage law. Studies show that "the percentage of youths unemployed increases as higher minimum wages are established...(Devils) are skillful in finding work...for idle hands...(We) create barbarians...The barbarians who finally sacked Rome...came from without. Much more efficient...will be the barbarians within" (173). Hammett notes that some of these unemployed have abdicated the work ethic, and live from day to day off the bottle, drugs, or theft. We have developed a pauper population, perpetually on relief (174).

The previous two paragraphs illustrate how law and technology are intertwined. The basic trouble rests with us--the people, and the professions. We should establish goals via democratic means in legislative hearings where knowledgeable professionals present pertinent testimony. It is here that engineers can describe the advantages and disadvantages of proposed plans, laws, or regulations. They can also develop policies on societal issues in their professional organizations. Statistically significant polls of the members could easily be conducted which reflect majority viewpoints and these could then be presented to law makers. Too often these policies are determined by the officers without much input from the engineering society membership, most of whom are not united enough to have much political influence.

There is a danger, however, in proposing that all groups like professional societies, manufacturers' associations, political action committees, and labor unions unite into separate lobbying groups. Their views should be considered if pertinent, but the lobbyists should be prevented from communicating orally with legislators except at Congressional hearings. They should be allowed to submit written reports at all times on other issues or goals which legislators can then study at their leisure. We ought to leave legislators alone, but hold them accountable for their judgement.

"All managers know the best way to run efficient organizations is to delegate authority and responsibility, and to hold subordinates accountable. If we, the people in

this Republic, delegate power to elected representatives, we should have the courtesy to
stop harassing them unless they request our
counsel. (Otherwise) participatory democracy
will talk itself to death and retire, if not
kill, the goose that lays the golden eggs...
There are subtle forces...endeavoring to
weaken what is left of our capitalistic
economy, (and to constrain its productive
capacity. The people involved misinterpret
its objectives, ignore its ability to generate wealth,) and endorse socialism over capitalism. One can only wonder if they have
even stood in an Algerian or Russian queue
wondering what goods may be available, shoddy
or not.

The sad truth is that in most of even
the free world the means of production is
publicly (government) owned or controlled.
These free nations and communistic countries
depend heavily upon others like the United
States for industrial output and technology
transfer where facilities are mostly privately owned. The data in the Table 8C, showing
this distribution for eighteen non-communist
nations, was taken from the (London)
ECONOMIST issue of December 30, 1978. Every
peasant in the world is aware of the correlation between U.S. productive capability and
our standard of living. It is a pity that
all public crusaders cannot also appreciate
the correlation between our privately owned
industrial capacity and technological supremacy. Although we use 30 percent of the world
resources, we have been surprisingly generous
in sharing that wealth" (175).

Much of the legislative harrassment by crusaders,
and of the legislative process, involves lawyers. Why
the United States needs twenty times as many lawyers
but only one-sixth as many scientists and engineers per
capita as Japan does is a mystery the reader can ponder. Table 8D, abstracted from various sources including the October 1976 issue of Harpers Magazine, shows
this data. It also indicates that the proportion of
lawyers in the United States doubled in 75 years. Formerly, in Colonial Massachusetts, courts allowed an
adult to plead his own case, or to retain someone else,
provided he gave him no fee or reward!

TABLE 8C - INDUSTRIES OWNED OR CONTROLLED
BY GOVERNMENT

	MAIL	SERVICE*	MFGR.**
Australia	100%	95%	0%
Austria	100	100	100
Belgium	100	70	87
Brazil	100	85	55
Britain	100	95	60
Canada	100	55	0
France	100	95	56
W. Germany	100	85	25
Holland	100	85	75
India	100	100	75
Italy	100	95	58
Japan	100	40	0
Mexico	100	90	80
S. Korea	100	55	25
Spain	100	65	44
Sweden	100	80	50
Switzerland	100	95	0
United States	100	10	0

* Service Industries (Airlines, Electricity,
Gas, Railroads, Telephone)
** Manufacturing Industries (Autos, Coal Oil,
Steel, Shipbuilding)

8D - PROFESSIONAL/POPULATION RATIO (175,181)

		U.S.	Israel	England	Japan
Lawyers	1900	1/1100	–	–	–
Lawyers	1925	1/700	–	–	–
Lawyers	1976	1/530	1/670	1/1600	1/10300
Accountants	1976	1/260	–	–	1/3100
Engineers & Scientists	1976	1/150	–	–	1/25

Since colonial times secular authority has been
entrusted to a legal elite. Perhaps this legal elite
has exceeded its critical mass. At least it seems so
for as the number of lawyers increase, so do the number
of laws passes, the enforcement needed, and court cases
pending. The pending legislation needed to accommodate
change should stress quality, not quantity. A fair
share of past legislation has emphasized quality and
favored free enterprise. Deregulation is a good example. However, it is the excess legislation which is

208

clogging our courts and which has caused Chief Justice Burger to warn the legal profession about its danger.

It is worth remembering that, as we establish our own societal goals, those areas of the world which have never tried the free capitalistic system remain in poverty and starvation. For instance, living standards in places like Singapore, Hong Kong, South Korea and Japan excell those in less developed countries even though the four cited have fewer natural resources to exploit.

It is essential to select goals so as to accommodate change, and not to constrain progress and productivity. For instance, it would have made no sense in the horse and buggy days to have taxed emerging automobile manufacturers to buy oats for government supported horses.

"During the next 25 years the world's population is expected to increase 55 percent. Whether the world's housing and supporting public infrastructure can be expanded at that pace, as resources dwindle and inflation continues, will pose a gigantic challenge, particularly for engineers. Their success will be accelerated by free market economics but retarded by governmental regulatory constraints. Further more, governmental funding for such national projects as the MX missile complex or alternate energy developments will divert resources from societal necessities. The practicality of such projects involves political trade-offs. Sometimes their completion produces economic benefits, as happened with the Panama Canal, Hoover Dam and space exploration. The interstate highway system benefited the trucking industry but bankrupted many railroads and delayed construction of urban subways. Other national projects like the French Maginot Line fortifications and the Egyptian pyramids proved useless except as memorials.

Engineers and scientists are generally better able than laymen to advise legislators on the practicality of such projects. Their recommendations will seldom be unanimous, and may consider alternate systems. Results of their studies should be presented at public hearings; acceptance of them may prevent costly blunders. An agency to advise Congress on such issues was initiated in the 1950's when what is now the Office of Technology Assessment was created (147). Some of its recommendations that involved the long term commitments needed for major projects did not fare too well. Politicians prefer to support short term solutions whose rapid completion enhances their bids for reelection... There will be a crucial need for engineering leadership in the future to help in selecting suitable public missions, in bolstering our free enterprise system, in curtailing unneeded governmental regulation, and in expanding our professional freedom for more responsible practice. Engineers (could)...assume active leadership roles which assist public bodies in the formulation of appropriate technical (goals) and societal missions. They are uniquely educated to formulate and to solve problems logically. Should they continue to remain as passive in the future as many of them have in the past, there is a grave danger that our fragile democratic environment will change, perhaps violently, into one far more dictatorial and collectivized" (76).

All professionals have a duty to assist in the formulation of goals and to dissent when necessary. The complexity of our technological civilization prompted many former engineering leaders to plead with their peers for greater civic involvement. Too few responded. They found technology more challenging. Employers seldom rewarded employees for civic involvement, nor did engineering programs focus enough on that obligation.

All professionals have a duty to engage in as much societal leadership as their talents allow. Their education and/or training should prepare them for that role. It is interesting to note the Communist commitment to leadership. Their novices are taught to lead, debate, write, organize, and infiltrate. Power flows to those who are active and who exercise responsibility.

The goals of engineering educators must likewise be directed toward grooming "their students not only to be technically competent at ever advancing levels, but to learn to exercise leadership in the industrial, gov-

210

ernmental and societal sectors (as well). Leaders and
ethical professionals will be badly needed worldwide to
preserve free enterprise and western civilization...
Like physicians, engineers of tomorrow will need a
sound understanding of their patient--society--if they
are to serve effectively...Like artists, society commissions engineers to fulfill a role. Both need talent, and both need training under stern disciplinari-
ans" (92).

Engineering educators and societies in the United
States began recently to sponsor about fifty fellowship
programs in societal leadership. The interns selected
as White House or Congressional Fellows serve as assistants in the offices of Senators and Congressmen, on
their Committees, or on support arms such as the Office
of Technology Assessment, gaining experience in the
political process through which societal missions are
formulated.

These fellowships are sponsored by such societies
as ASCE, ASME, the American Association of the Advancement of Science (AAAS), or the University of Washington
with its Washington Internship for Students of Engineering (WISE) program. At least a start has been
made.

Societal goals become increasingly important as
our technological civilization becomes so critically
interdependent that war should be unthinkable. Such
goals should relate to the optimal distribution of necessities like nutritious food, strategic materials and
essential technology. Their distribution would be enhanced if only nations could accept free trade, unre-
stricted emigration and selective immigration. The
reader may argue that these are essentially political
ideals. And so they are. But they are also hopelessly
entangled with economics and technology. That is why
it is so essential that all knowledgeable professionals
help to establish these goals.

It is fruitless to perpetuate the mutual incompre-
hension that polarizes the literary and scientific in-
tellectuals C. P. Snow described in THE TWO CULTURES.
Nor is it helpful for any elitists to accept science
but exclude its application. Brown (11), it may be
remembered (Article 1A), was convinced that engineers
were the best prepared to act as the moderators between
the two worlds of science and the humanities. All professionals should be involved in at least setting the

211

goals. Simon Watt, in commenting on the 1979 Finniston Report, which described the failure of British engineers to continue Britain's industrial leadership after 1880, felt that "It may be necessary for engineers to...actively challenge the intellectual position of the elite...through professional engineering institutions, or at least to take part in movements of dissent" (163).

Society's goals are ill chosen when companies find it easier to increase profits through mergers and marketing, to relegate new and better products to a secondary goal, and by these mergers to reduce the taxes they pay to operate their government. What is worse, they may feel justified even though production is curtailed and employment decreased.

Man has in the past and will in the future formulate goals and find political, moral, and technical solutions to his problems. These solutions, and the goals from which they are generated, would be more appropriate if the leadership were more enlightened. It is the leaders in the past who have literally been their brother's keeper, who governed their fellowmen, and who built industries to provide them with jobs, subsistence, and their luxuries as well. Now, as society becomes evermore complex, they must "try harder."

The interdependence of all people, all professionals and all leaders on each other is best illustrated by the following true story which Judge Ted Dalton of the Federal District Court of the Western District of Virginia likes to tell.

"If I were to name one unchanging value in this changing world above all others, it would be that of help and service. One can illustrate it no better than by repeating my favorite story, and to urge you that you remember it and that you keep it as your philosophy of life.

About one hundred years ago a humble Scotsman, strolling near his home in Darvel, rescued a boy who was dangerously mired in a bog. The lad turned out to be the son of a nobleman--Lord Marlborough, in fact, who as you know was the father of Sir Winston Churchill. The Scotsman refused Marlborough's gratitude for himself but

212

agreed that Marlborough might help him educate his son. The Lord Marlborough did.

The boy ultimately was graduated from
St. Mary's Hospital Medical School. His
Name? Fleming - Sir Alexander Fleming, the
discoverer of penicillin.

There is a sequel. During World War II,
when Britain faced its darkest hour of
crisis, Marlborough's son, the then Prime
Minister of Great Britain, was stricken with
pneumonia. He lived because of penicillin.
His Name? Winston Churchill."

It is we who can manage ideas and lead people or
formulate goals, so as to leave the world a bit better
than we found it, who must never lose hope. Nor must
we ever forget that as technology shrinks the time and
space frames in which the world functions, the challenges expand. Man now possesses the knowledge to enhance society's cultural atmosphere and living standards by orders of magnitude, or to annihilate everything.

"So little done.
So much to do."
So little time
to do it in.

And yet, even though "we go around only once," there
will be ample time for those engineers who care enough
to lead and to serve the engineering profession and its
emerging public purpose.

"Ut prosimus et ducamus.
That we may serve and lead." (116)

213

Appendix A
ABET Engineering Society
Membership

ENGINEERING SOCIETY MEMBERSHIP

IN THE

ACCREDITATION BOARD FOR ENGINEERING AND TECHNOLOGY

PARTICIPATING BODIES

AAEE	American Academy of Environmental Engineers
ACSM	American Congress on Surveying and Mapping
AIAA	American Institute of Aeronautics and Astronautics, Inc.
AIChE	American Institute of Chemical Engineers
AIIE	American Institute of Industrial Engineers
AIME	American Institute of Mining, Metallurgical and Petroleum Engineers
ANS	American Nuclear Society
ASAE	American Society of Agricultural Engineers
ASCE	American Society of Civil Engineers
ASEE	American Society for Engineering Education
ASHRAE	American Society of Heating, Refrigerating and Air-Conditioning Engineers, Inc.
ASME	American Society of Mechanical Engineers
IEEE	The Institute of Electrical and Electronic Engineers, Inc.
NCEE	National Council of Engineering Examiners
NICE	National Institute of Ceramic Engineers
NSPE	National Society of Professional Engineers
SAE	Society of Automotive Engineers
SME	Society of Manufacturing Engineers
SNAME	Society of Naval Architects and Marine Engineers

MEMBER BODIES

ASM	American Society for Metals

Appendix B
ASCE Code of Ethics
and Guidelines

CODE OF ETHICS*

Effective January 1, 1977

FUNDAMENTAL PRINCIPLES**

Engineers uphold and advance the integrity, honor and dignity of the engineering profession by:

1. using their knowledge and skill for the enhancement of human welfare;

2. being honest and impartial and serving with fidelity the public, their employers and clients;

3. striving to increase the competence and prestige of the engineering profession; and

4. supporting the professional and technical societies of their disciplines.

FUNDAMENTAL CANONS

1. Engineers shall hold paramount the safety, health and welfare of the public in the performance of their professional duties.

2. Engineers shall perform services only in areas of their competence.

3. Engineers shall issue public statements only in an objective and truthful manner.

4. Engineers shall act in professional matters for each employer or client as faithful agents or trustees, and shall avoid conflicts of interest.

5. Engineers shall build their professional reputation on the merit of their services and shall not compete unfairly with others.

6. Engineers shall act in such a manner as to uphold and enhance the honor, integrity, and dignity of the engineering profession.

7. Engineers shall continue their professional development throughout their careers, and shall provide opportunities for the professional development of those engineers under their supervision.

---

*As adopted September 25, 1976.
**The American Society of Civil Engineers adopted THE FUNDAMENTAL PRINCIPLES of the ECPD Code of Ethics of Engineers as accepted by the Engineers' Council for Professional Development (ECPD).
(By ASCE Board of Direction action April 12-14, 1975)

216

ASCE GUIDELINES TO PRACTICE UNDER THE FUNDAMENTAL CANONS OF ETHICS

CANON 1. Engineers shall hold paramount the safety, health and welfare of the public in the performance of their professional duties.

a. Engineers shall recognize that the lives, safety, health and welfare of the general public are dependent upon engineering judgments, decisions and practices incorporated into structures, machines, products, processes and devices.

b. Engineers shall approve or seal only those design documents, reviewed or prepared by them, which are determined to be safe for public health and welfare in conformity with accepted engineering standards.

c. Engineers whose professional judgment is overruled under circumstances where the safety, health and welfare of the public are endangered, shall inform their clients or employers of the possible consequences.

d. Engineers who have knowledge or reason to believe that another person or firm may be in violation of any of the provisions of Canon 1 shall present such information to the proper authority in writing and shall cooperate with the proper authority, in furnishing such further information or assistance as may be required.

e. Engineers should seek opportunities to be of constructive service in civic affairs and work for the advancement of the safety, health and well-being of their communities.

f. Engineers should be committed to improving the environment to enhance the quality of life.

CANON 2. Engineers shall perform services only in areas of their competence.

a. Engineers shall undertake to perform engineering assignments only when qualified by education or experience in the technical field of engineering involved.

b. Engineers may accept an assignment requiring education or experience outside of their own fields of competence, provided their services are restricted to those phases of the project in which they are qualified. All other phases of such project shall be performed by qualified associates, consultants, or employees.

c. Engineers shall not affix their signatures or seals to any engineering plan or document dealing with subject matter in which they lack competence by virtue of education or experience or to any such plan or document not reviewed or prepared under their supervisory control.

CANON 3. Engineers shall issue public statements only in an objective and truthful manner.

a. Engineers should endeavor to extend the public knowledge of engineering, and shall not participate in the dissemination of untrue, unfair or exaggerated statements regarding engineering.

b. Engineers shall be objective and truthful in professional reports, statements, or testimony. They shall include all relevant and pertinent information in such reports, statements, or testimony.

c. Engineers, when serving as expert witnesses, shall express an engineering opinion only when it is founded upon adequate knowledge of the facts, upon a background of technical competence, and upon honest conviction.

d. Engineers shall issue no statements, criticisms, or arguments on engineering matters which are inspired or paid for by interested parties, unless they indicate on whose behalf the statements are made.

e. Engineers shall be dignified and modest in explaining their work and merit, and will avoid any act tending to promote their own interests at the expense of the integrity, honor and dignity of the profession.

CANON 4. Engineers shall act in professional matters for each employer or client as faithful agents or trustees, and shall avoid conflicts of interest.

a. Engineers shall avoid all known or potential conflicts of interest with their employers or

217

clients and shall promptly inform their employers or clients of any business association, interests, or circumstances which could influence their judgment or the quality of their services.

b. Engineers shall not accept compensation from more than one party for services on the same project, or for services pertaining to the same project, unless the circumstances are fully disclosed to and agreed to, by all interested parties.

c. Engineers shall not solicit or accept gratuities, directly or indirectly, from contractors, their agents, or other parties dealing with their clients or employers in connection with work for which they are responsible.

d. Engineers in public service as members, advisors, or employees of a governmental body or department shall not participate in considerations or actions with respect to services solicited or provided by them or their organization in private or public engineering practice.

e. Engineers shall advise their employers or clients when, as a result of their studies, they believe a project will not be successful.

f. Engineers shall not use confidential information coming to them in the course of their assignments as a means of making personal profit if such action is adverse to the interests of their clients, employers or the public.

g. Engineers shall not accept professional employment outside of their regular work or interest without the knowledge of their employers.

CANON 5. Engineers shall build their professional reputation on the merit of their services and shall not compete unfairly with others.

a. Engineers shall not give, solicit or receive either directly or indirectly, any commission, political contribution, or a gift or other consideration in order to secure work, exclusive of securing salaried positions through employment agencies.

b. Engineers should negotiate contracts for professional services fairly and on the basis of demonstrated competence and qualifications for the type of professional service required.

c. Engineers shall not request, propose or accept professional commissions on a contingent basis under circumstances in which their professional judgments may be compromised.

d. Engineers shall not falsify or permit misrepresentation of their academic or professional qualifications or experience.

e. Engineers shall give proper credit for engineering work to those to whom credit is due, and recognize the proprietary interests of others. Whenever possible, they shall name the person or persons who may be responsible for designs, inventions, writings or other accomplishments.

f. Engineers may advertise professional services in a way that does not contain self-laudatory or misleading language or is in any other manner derogatory to the dignity of the profession. Examples of permissible advertising are as follows:

Professional cards in recognized, dignified publications, and listings in rosters or directories published by responsible organizations, provided that the cards or listings are consistent in size and content and are in a section of the publication regularly devoted to such professional cards.

Brochures which factually describe experience, facilities, personnel and capacity to render service, providing they are not misleading with respect to the engineer's participation in projects described.

Display advertising in recognized dignified business and professional publications, providing it is factual, contains no laudatory expressions or implication and is not misleading with respect to the engineer's extent of participation in projects described.

A statement of the engineers' names or the name of the firm and statement of the type of service posted on projects for which they render services.

218

Preparation or authorization of descriptive articles for the lay or technical press, which are factual, dignified and free from laudatory implications. Such articles shall not imply anything more than direct participation in the project described.

Permission by engineers for their names to be used in commercial advertisements, such as may be published by contractors, material suppliers, etc., only by means of a modest, dignified notation acknowledging the engineers' participation in the project described. Such permission shall not include public endorsement of proprietary products.

g. Engineers shall not maliciously or falsely, directly or indirectly, injure the professional reputation, prospects, practice or employment of another engineer or indiscriminately criticize another's work.

h. Engineers shall not use equipment, supplies, laboratory or office facilities of their employers to carry on outside private practice without the consent of their employers.

CANON 6. Engineers shall act in such a manner as to uphold and enhance the honor, integrity, and dignity of the engineering profession.

a. Engineers shall not knowingly act in a manner which will be derogatory to the honor, integrity, or dignity of the engineering profession or knowingly engage in business or professional practices of a fraudulent, dishonest or unethical nature.

CANON 7. Engineers shall continue their professional development throughout their careers, and shall provide opportunities for the professional development of those engineers under their supervision.

a. Engineers should keep current in their specialty fields by engaging in professional practice, participating in continuing education courses, reading in the technical literature, and attending professional meetings and seminars.

b. Engineers should encourage their engineering employees to become registered at the earliest possible date.

c. Engineers should encourage engineering employees to attend and present papers at professional and technical society meetings.

d. Engineers shall uphold the principle of mutually satisfying relationships between employers and employees with respect to terms of employment including professional grade descriptions, salary ranges, and fringe benefits.

219

Obligation Of An Engineer

I am an Engineer. In my profession I take deep pride. To it I owe solemn obligations.

Since the Stone Age, human progress has been spurred by the engineering genius. Engineers have made usable Nature's vast resources of material and energy for Mankind's benefit. Engineers have vitalized and turned to practical use the principles of science and the means of technology. Were it not for this heritage of accumulated experience, my efforts would be feeble.

As an Engineer, I

_____

pledge to practice integrity and fair dealing, tolerance and respect; and to uphold devotion to the standards and the dignity of my profession, conscious always that my skill carries with it the obligation to serve humanity by making the best use of Earth's precious wealth.

As an Engineer, in humility and with the need for Divine Guidance, I shall participate in none but honest enterprises. When needed, my skill and knowledge shall be given without reservation for the public good. In the performance of duty and in fidelity to my profession, I shall give the utmost.

Signature
_____

Date
_____

Location
_____

THE ORDER OF THE ENGINEER

The "Order of the Engineer" is a fellowship of engineers who are trained in science and technology and dedicated to the practice, teaching or administration of their profession.

Initiation into the Order includes acceptance of the "Obligation", and a stainless steel ring to be worn on the little finger of the working hand. Only those who have met the high standards of professional engineering training or experience are invited to accept the Obligation, which is voluntarily received for life. This commitment is not a trivial act but is, rather, like the wedding of the engineer with his profession. The ring is worn as a visible symbol to attest to the wearer's calling and symbolizes the unity of the profession in its goal of benefiting mankind. The stainless steel from which the ring is made depicts the strength of the profession.

The Order was originated by several members of the Ohio Society of Professional Engineers who were inspired by The Ritual of the Calling of an Engineer, a Canadian organization whose adherents wear an iron ring. The first ceremony of the Order was at Cleveland State University on June 4, 1970.

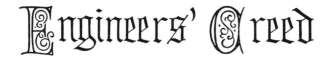

Engineers' Creed

As a Professional Engineer, I dedicate my professional knowledge and skill to the advancement and betterment of human welfare.

I pledge:

To give the utmost of performance;

To participate in none but honest enterprise;

To live and work according to the laws of man and the highest standards of professional conduct;

To place service before profit, the honor and standing of the profession before personal advantage, and the public welfare above all other considerations.

In humility and with need for Divine Guidance, I make this pledge.

Adopted by
National Society of Professional Engineers
June, 1954

Engineers' Council for Professional Development

FAITH OF THE ENGINEER

I AM AN ENGINEER. *In my profession I take deep pride, but without vainglory; to it I owe solemn obligations that I am eager to fulfill.*

As an Engineer, I will participate in none but honest enterprise. To him that has engaged my services, as employer or client, I will give the utmost of performance and fidelity.

When needed, my skill and knowledge shall be given without reservation for the public good. From special capacity springs the obligation to use it well in the service of humanity; and I accept the challenge that this implies.

Jealous of the high repute of my calling, I will strive to protect the interests and the good name of any engineer that I know to be deserving; but I will not shrink, should duty dictate, from disclosing the truth regarding anyone that, by unscrupulous act, has shown himself unworthy of the profession.

Since the Age of Stone, human progress has been conditioned by the genius of my professional forbears. By them have been rendered usable to mankind Nature's vast resources of material and energy. By them have been vitalized and turned to practical account the principles of science and the revelations of technology. Except for this heritage of accumulated experience, my efforts would be feeble. I dedicate myself to the dissemination of engineering knowledge, and, especially to the instruction of younger members of my profession in all its arts and traditions.

To my fellows I pledge, in the same full measure I ask of them, integrity and fair dealing, tolerance and respect, and devotion to the standards and the dignity of our profession; with the consciousness, always, that our special expertness carries with it the obligation to serve humanity with complete sincerity.

Prepared by the Ethics Committee

References

NOTE: Pages on which references are cited are listed at
          the end of each reference as c-p; where p = page no.
          Example Reference 1 is cited (c-9) on page 9.

1. Preliminary Report, Committee on Society Objectives, Planning
    and Objectives, October 1972, "Goals for ASCE," *Civil Engi-
    neering*, Feb. 1973, p. 50-53: c-9.
2. Dillard, Joseph K., "Professional Schools of Engineering - A
    Call for Action," IEEE *Transactions on Education*, Vol. E-
    19, No. 4, Nov. 1976: c-12.
3. Dixon, M. D., "But Are You Really A Professional," *Engineer*,
    Engineers Joint Council, Jan.-Feb., 1968: c-7.
4. Florman, S. C., "The Existential Pleasures of Engineering,"
    St. Martins Press, New York, 1976: c-2,188.
5. Lukasiewicz, J., "The Railway Game," *McClelland and Steward,
    Ltd.*, Toronto, Canada, 1976, p. 218-225: c-10.
6. Pletta, Dan H., discussion of "Report of Observations of At-
    tendees at the Engineering Foundation Conference on Life
    Span of Structures," M. H. Salgo, Chairman, Performance of
    Structures Research Council, ASCE *Engineering Issues*, Jan.
    1976, p. 52-54: c-11.
7. Pletta, Dan H., "Social Science Emphasis in Engineering Edu-
    cation," ASCE *Engineering Issues*, Oct. 1975, p. 509-519: c-
    12,26,128,129,154.
8. Pletta, Dan H., "Tomorrow's Professional Engineer--The Pub-
    lic's Advocate," ASCE *Engineering Issues*, July 1977, p.
    223-231: c-12, 143.
9. Pound, Roscoe, "The Lawyer from Antiquity to Modern Times,"
    West Publishing Co., St. Paul, MN, 1953, p. 5: c-5.
10. Whitelaw, Robert L., "The Professional Status of the American
    Engineer: A Bill of Rights," NSPE *Professional Engineer*,
    August 1975: c-13,36,60.
11. Brown, J. Douglas, "The Role of Engineering as a Learned Pro-
    fession," *Princeton University*, 11 Oct. 1962: c-3,186, 212.
12. NSPE *Report*, "Qualifications for Practice and Leadership of
    Engineering Professionals," Registration and Qualifications
    for Practice Committee, July 1980: c-6,16,39,41,83-86,94.

13. Flexner, Abraham, "Universities," *Oxford University Press*, New York, 1967, p. 41-42; 1930, p. 29-30: c-6.
14. Wisely, William H., "Public Obligation and the Ethics System," ASCE *Issues in Engineering*, July 1979, p. 131-135: c-6.
15. Shaw, George Bernard, "The Doctor's Dilemma," Act I. *London Constable & Co.*, Ltd., 1913: c-15.
16. Nash, Ogden, "I Yield to My Learned Brother, or Is There a Candlestick Maker in the House?" *The Primrose Path*, Simon and Schuster, New York, 1935: c-15.
17. Pletta, Dan H., "Personally Speaking: Professional Freedom for Responsible Practice," *Mechanical Engineering*, Oct. 1978: c-17,104,106,109.
18. Rabinow, Jacob, "Is American Genius Being Stifled?" *U.S. News and World Report*, 23 Dec. 1974: c-18,91.
19. Spencer, Richard, "Editorial," *Saturday Evening Post*, 25 May 1957: c-18.
20. Special Report, "A Vanishing American--The Small U.S. Inventor," *U.S. News and World Report*, 23 Nov. 1956: c-18.
21. Barzun, Jacques, "The Professions Under Siege," *Harpers*, Oct. 1978: c-19,163,184.
22. Pletta, Dan H., "Encouraging Professional Development with Engineering Specialty Boards, Diplomates and Unity," *Proceedings ASEE Southeastern Section Conference*, April 1979: c-19,73,76,96,97.
23. Pletta, Dan H., "Guest Comment," Apr. 1978 and "Reader Forum," Aug. 1978, *Professional Engineer*: c-20,23.
24. Schock, E. P., "Eight Surprises," *Louisiana Engineer*, Oct. 1976: c-24.
25. "Less Time, More Options," *Report on Education Beyond the High School, Carnegie Commission on Higher Education*, Jan. 1971: c-27.
26. Hollister, S. C., "Engineers Must Be Upgraded to Solve the Manpower Shortage," *Civil Engineering*, Sept. 1952, p. 82-83: c-27.
27. Pletta, Dan H. and George A. Gray, "To Practice or Perish," ASCE *Engineering Issues*, Oct. 1971, p. 131-146: c-28,80, 130.
28. Rudoff, A. and D. Lucken, "The Engineer and His Work: A Sociological Perspective," *Science*, Vol. 172, June 11, 1971, p. 1103: c-29,114.
29. Lukasiewicz, J., "The Ignorance Explosion: A Contribution to the Study of Confrontation of Man with the Complexity of Science-Based Society and Environment," *Trans. New York Academy of Sciences*, Series II, Vol. 34, May 1972, p. 373: c-29,30.
30. *Position paper* on "Sizes and Weights of Vehicles" prepared by the Structures Group Executive Committee, Metropolitan Section, *American Society of Civil Engineers*, Oct. 1974: c-31.

31. Pletta, Dan H. and George A. Gray, "Engineering Professionalism--A Light at the End of the Tunnel," ASCE *Engineering Issues*, Apr. 1977, p. 157-163: c-34.
32. Toffler, A., "Future Shock," *Random House*, New York, NY, 1970: c-34,186.
33. Burkett, W., "The Consumer Product Safety Commission," *New Engineer*, Dec. 1974: c-35.
34. Manne, H. G. and H. C. Wallich, "The Modern Corporation and Social Responsibility," *American Enterprise Institute for Public Policy Research*, Washington, D.C., 1972: c-36.
35. Salvadori, M. G., "Social Concern for Engineering Education," *Civil Engineering*, June 1974, p. 70-73: c-37.
36. "Engineering for the Benefit of Mankind," *Conference Proceedings*, National Academy of Engineering, 1967: c-37.
37. Forrester, J. W., "Engineering and Engineering Practice in the Year 2000," *Conference Proceedings*, National Academy of Engineering, 1967: c-36.
38. Beakley, George C. and H. W. Leach, "Engineering: An Introduction to a Creative Profession," *Macmillan & Co.*, New York, NY, 1967: c-43.
39. Mantell, Murray I., "Ethics and Professionalism in Engineering," *Macmillan & Co.*, New York, NY, 1964: c-43.
40. "History of Technology," *Encyclopedia Britannica*: c-44,45.
41. Dallaire, Gene, "The Engineer: What Role in the Development of Civilization?," *Civil Engineering*, Oct. 1977, p. 65-74: c-61,62,190.
42. Rougier, Louis, "The Genius of the West," *Nash Publishing*, Los Angeles, CA, 1971: c-49-52.
43. Hazlitt, Henry, "The Conquest of Poverty," *Arlington House*, New Rochelle, NY, 1976: c-50.
44. Hill, Ivan, "The Ethical Basis of Economic Freedom," *American Viewpoint, Inc.*, Chapel Hill, NC, 1976: c-51,163,171.
45. Rifkin, Jeremy and Ted Howard, "Entropy," *Viking Press, Inc.*, New York, NY, 1980: c-54,95,101-103.
46. Kondratieff, Nikolai D., "The Long Waves of Economic Life," *Archiv fur Sozialwissenschaft und Sozialpolitik*, 1926; translated by W. F. Stolper in *Review of Economic Statistics*, Nov. 1935: c-55-57.
47. Snyder, Julian M., "The Economic Long Wave: Key to Your Investment Survival," *International Moneyline*, Darien, CT, 1982: c-55-57.
48. Schumpeter, Joseph, "Business Cycles," *McGraw Hill Book Co.*, New York, NY, 1939: c-55,57.
49. Mensch, Gerhard, "Stalemate in Technology: Innovations Overcome the Depression," *Ballinger Pub. Co.*, Cambridge, MA, 1979: c-55.
50. Hall, Peter, "The Next Economic Boom," *World Press Review*, June 1981: c-55,168,173.
51. Ogburn, W. F., "Social Change," *Viking Press, Inc.*, New York, NY, 1950: c-55-56.

52. Clemence, R. V. and Francis S. Doody, "The Schumpeterian System," *Addison-Wesley Press, Inc.*, Cambridge, MA, 1950: c-58.

53. Heilbroner, R. L., "In the Name of Profit: Profiles in Corporate Responsibility," *Warner Paper Back Library*, No. 0-446-78178-9: c-59,87,89,172,180.

54. Nader, Ralph, Peter Petkas and Kate Blackwell, "Whistle Blowing: Profiles in Conscience and Courage," *Bantam Books*, No. 553-07645-195: c-53,87,89,172,176,180.

55. Eddy, Paul, Elaine Potter and Bruce Page, "Destination Disaster," *NY Times Book Co.*, 1974: c-53,87,89,172.

56. Veblen, Thorstein, "The Theory of the Leisure Class," *Macmillan & Co., Ltd.*, London, 1899: c-62.

57. Wright, J. Patrick, "On a Clear Day You Can See General Motors," *Avon Books*, New York, NY, 1979, No. ISBM 0-380-51722-1. (John Z. DeLorean's look inside the automotive giant). c-62.

58. Layton, Edwin T., Jr., "The Revolt of the Engineers," *Case Western University Press*, Cleveland, OH, 1971: c-63,64,70.

59. Behr, Peter, "Executives Take Heat for Business Decline," LA Times-Washington Post Service, *Roanoke (VA) Times and World News*, Jan. 31, 1982: c-191.

60. Pletta, Dan H., "Why Professional Schools for Engineers?," ASCE *Issues in Engineering*, Oct. 1980, p. 349-364: c-15,69,70,71,73,75,134,135,139,152-155,163,172.

61. Green, Mark, "A Question of Leadership," LA-Times Washington Post Service, *Roanoke (VA) Times and World News*, May 24, 1981: c-192.

62. Pletta, Dan H., "Ingenieur Diplomates--Public Advocates Worldwide," ASCE *Specialty Conference on Ethics, Professionalism and Maintaining Competence*, Columbus, OH, Mar. 10-11, 1977, p. 20-26: c-62,73,75,94,170.

63. Wisely, W. H., "The American Civil Engineer," *The American Society of Civil Engineers*, New York, NY, 1974: c-77,78.

64. Pletta, Dan H., "The Development of Professional Schools," NSPE *Workshop on Professional Schools of Engineering*, Mar. 13-14, 1978: c-80,117.

65. Pletta, Dan H., "Guest Editorial--Consulting and Licensing of Engineering Faculties: Sparking a Professional Esprit de Corps," ASEE *Engineering Education*, Nov. 1976: c-133,159.

66. Wickenden, William E., "The Second Mile," reprinted in ASEE *Engineering Education*, April 1981. First presented in 1941 before the Engineering Institute of Canada: c-87.

67. *Intersociety Conference* on Engineering Ethics, May 1975, p. 82-83: see Ref. 139 also: c-89.

68. Zuesse, Eric, "Love Canal: The Truth Seeps Out," *Reason*, Feb. 1981: c-89.

69. Reed, Adam V., "Who Caused Three Mile Island," *Reason*, Aug. 1980: c-89.

70. Editorial, "Who's Threatened Most?" *The Economist*, Aug. 22, 1981: c-92.

226

71. Markey, Howard T., "Risk Management: A Congressional Chore," *Virginia Engineer*, Aug. 1981: c-94.
72. Mintz, Morton, "Justice Asks New Patent Procedures," *Washington Post*, Dec. 4, 1972: c-97.
73. Rabinow, Jacob, Testimony Before the Subcommittee on Domestic and International Scientific Planning and Analysis of the Committee on Science and Technology, Congress of the United States, April 29, 1976: c-97.
74. "Taking the Shackles Off American Industry," *U.S. News & World Report*, Washington, D.C., Nov. 12, 1979: c-97.
75. Schumacher, E. F., "Small Is Beautiful: Economics as if People Mattered," Perennial Library, *Harper & Row Publisher, Inc.*, New York, NY, 1973: c-97.
76. Pletta, D. H., "Engineering: Answers for the 1980's," ASCE *Issues in Engineering*, July 1981, p. 193-204: c-97,104,210.
77. Farb, Peter, "Humankind," *Houghton-Mifflin*, Boston, 1978, p. 181-182: c-99.
78. McKetta, John J., "The U.S. Energy Problem Grows Worse and Worse and...," Bulletin, *National Council for Environmental Balance*, Louisville, KY, 1980: c-100,104,192,210.
79. "Trends," *Reason*, Sept. 1981: c-99,100.
80. Hubbard, Barbara M., "Critical Path to an All Win World: Bucky Fuller Designs the New Age," *The Futurist*, June 1981: c-102-104.
81. Oparin, A. I., "The Origin of Life," *MacMillan Press*, New York, NY, 1938: c-103.
82. McMaster, R. E., Jr., "Cycles of War—The Next Six Years," *Timberline Trust*, Kalispell, MT, 59901, 1977: c-104.
83. Housman, A. E., "Last Poems," *Grant Richards, Ltd.*, London, England, 1922, p. 28: c-104.
84. Hazlett, Thomas Winslow, "Big Business Comes Out of Its Corner Fighting," *Reason*, July 1977: c-105.
85. Wilkie, Wendell L., "One World," *University of Illinois Press*, Urbana, IL, 1966: c-105,106.
86. Chamberlain, John, a book review of "The Future of Business Regulation: Private Action and Public Demand," *Amacon*, New York, NY, 1979: c-108.
87. Klein, Stanley, "Will Engineers Unionize," *New Engineer*, Nov. 1972: c-113.
88. Research Report, "Manager Unions," *American Management Association, Inc.*, New York, NY: c-113.
89. Pletta, Dan H., "Pressures in Engineering Education," NSPE *Professional Engineer*, Feb. 1973: c-113,114,144,145,156, 157.
90. Pletta, Dan H., "Today's Challenge to Engineers: Lead or be Led," ASEE *Engineering Education*, April 1981: c-118,187.
91. Grayson, Lawrence P., "A Brief History of Engineering Education in the United States," ASEE *Engineering Education*, Dec. 1977: c-125.
92. NSPE *Report*, "Long Range Strategy for NSPE Involvement in ECPD," Aug. 1976: c-133,159,211.

93. "Final Report of the National Commission on Product Safety," June 1970, *Library of Congress*, No. 76-600753: c-115.
94. "The President's Report on Occupational Safety and Welfare," May 1972, U.S. Depository Publication No. 14173: c-115.
95. Weinstein, A. S. and R. H. Piehler (engineers); and A. D. Twerski and W. A. Donaher (attorneys): "Product Liability; An Inter-action of Law and Technology," *Duquesne Law Review*, Vol. 12, No. 3, Spring 1974: c-115.
96. NSPE *Report* "Challenge of the Future: Professional Schools of Engineering," Professional Schools Task Force, 1976: c-128,131,133.
97. Wickenden, W. E., "Report of the Investigation of Engineering Education (1923-29)," *Soc. for the Promotion of Engineering Education*, Vol. I, 1930, Vol. II, 1934: c-130.
98. Hammond, H. P., "Aims and Scope of Engineering Education," *Journal of Engineering Education*, Vol. 30, No. 7, Mar. 1940; "Engineering Education After the War," *Journal of Engineering Education*, Vol. 34, No. 9, May 1944: c-130.
99. Grinter, L. E., "Report on Evaluation of Engineering Education (1952-55)," *Journal of Engineering Education*, Vol. 46, No. 1, Sept. 1955: c-130.
100. Walker, W. A., "Goals of Engineering Education," ASEE *Final Report*, Jan. 1968: c-130.
101. AAES *Report*, "Engineering Manpower Bulletin," No. 65, Dec. 1982: c-130,131.
102. Everitt, W. D., "The Phoenix--A Challenge to Engineering Education, IEEE *Transactions on Education*, Sept. 1944 and Nov. 1980: c-130.
103. Rodenberger, Charles A., "The School of Professional Engineering: An Administrative Model," ASEE *Engineering Education*, Apr. 1981: c-131.
104. Pletta, Dan H. and George A. Gray. "Engineering Accountability vs. Corporate Responsibility," *Proceedings* IIT/NSF 2nd National Conference on Ethics in Engineering, Mar. 1982: c-27,106.
105. Feld, Jacob, "Construction Failure," *John Wiley & Sons*, Inc., 1968: c-106,107.
106. Pletta, Dan H., "A New Approach for the Doctor's Degree for Engineers," *Journal of Engineering Education*, Nov. 1950: c-133.
107. Bjorhovde, Reidar, "Engineering Study in the U.S. and Europe," *Engineering Education*, Nov. 1981: c-136.
108. Samson, Charles H., Jr., and Dan H. Pletta, "A New Educational Environment for Tomorrow's Professional Engineer," ASEE *Frontiers in Education Conference*, 1978: c-137.
109. Pletta, Dan H., "Society's Needs: Leaders, Learned Professions and Ombudsmen," NSPE *Workshop* on Leadership and Professionalism in Engineering Education, Nov. 1977: c-138, 170.
110. Proctor, Carlton S., "Professional Aims of the Civil Engineer," *Civil Engineering*, Mar. 1939, p. 151-152: c-139.

111. Rose, David J., "Continuity and Change: Thinking New Ways About Large and Persistent Problems," *Technology Review*, Feb./Mar. 1981: c-142.

112. Seamans, Robert F. and Kent F. Hansen, "Engineering Education for the Future," *Technology Review*, Feb./Mar. 1981: c-142.

113. Lippmann, Walter, "Education vs. Western Civilization," *American Scholar*, Spring 1941. Presented at the 1940 Annual AAAS Meeting: c-140.

114. Pletta, Dan H., "Are We Really Educating Engineering Professionals?" *Civil Engineering*, Oct. 1980, p. 102: c-140.

115. "Report of the Task Committee on Professional Education," Adolph J. Ackerman (Chrmn.), *Civil Engineering*, Feb. 1958, p. 111: c-135.

116. Pletta, Dan H, "Ut Prosimus et Ducamus: That We May Serve and Lead," ASCE *Civil Engineering Education*, Vol. 1, Part 1, 1974, p. 459-468: c-73,135,213.

117. "Civil Engineering Education," *Conference Report* sponsored by ASCE, ASEE, Cooper Union and NSF, 1960, p. 137-151: c-136,158,159.

118. "Balloting on the 1960 Conference on Civil Engineering Education Reported," *Civil Engineering*, July 1961, p. 43: c-136.

119. Dauffenbach, Robert, "Forecasting Engineering Manpower Supply on Inter-Occupational Mobility," presented at the *Engineering Foundation Conference*, Franklin Pierce College, Aug. 1978: c-146-147.

120. Pletta, Dan H. and George A. Gray, "Engineering Leadership and Technological Teamwork: Preserver of Free Enterprise," ASCE *Conference on Civil Engineering Education*, Apr. 1979, Vol. 1, p. 271-280: c-146,148,149.

121. *PER Special Reports*, "The American Management Associations: The Emerging International Business-Grant School," Vol. 1, No. 1, Fairfax, VA: c-149.

122. Sinclair, George, "Guest Edutorial: The Decline of Professionalism in Engineering," IEEE *Transactions in Education*, Nov. 1980: c-149.

123. Brown, Stuart M. and Walter R. Lynn, "Commentary on Professionalism and Specialization in Engineering Education," *Engineering-Cornell Quarterly*, Spring 1981: c-150.

124. Lynn, Walter R., "Engineering and Society Programs in Engineering Education," *Science*, Jan. 1977: c-151.

125. "Science and Engineering Education for the 1980's and Beyond," *NSF/Department of Education Report*," Oct. 1980: c-152.

126. Matheny, Charles W., Jr., "Creation of a State School of Civil Engineering for Florida," *Florida Engineering Society Jour.*, Oct. 1978: c-144,152.

127. Gloyna, Earnest, "The Cost of Engineering Education at the University of Texas/Austin," presented at the NSPE/PEE *Conference on Cost of Providing Engineering Education*,

Jan. 1979: c-153.

128. Berg, Ivar, "Education and Jobs: The Great Training Robbery," *Praeger Publishers*, New York, NY, 1970: c-154.

129. Batdorf, S. B., "On Mechanics and Society," presented at a meeting of the ASME Applied Mechanics Division, June 1971: c-154.

130. Tribus, M., "Technology for Tomorrow," *Mechanical Engineering*, May 1971: c-155.

131. *Address* by Earl Warren at the Louis Marshall Award Dinner of the Jewish Theological Seminary of America, New York, NY, Nov. 12, 1962: c-162.

132. Ackerman, Adolph J., "Slow Death of a Free Profession," IEEE *Transactions on Aerospace and Electronic Systems*, May 1971: c-162.

133. Ladensen, Robert F., "Professional Codes of Ethics: Some Questions," *Perspectives*, IIT Center for the Study of Ethics in the Professions, Mar. 1981: c-165.

134. Hall, Peter, "It's Still a Tangled Ethical Web," *Graduating Engineer*, Sept. 1981: c-168,173.

135. "National Society of Professional Engineers v. United States," *Supreme Court of the United States. No. 76-1767*, April 1978: c-168.

136. Florman, Samuel, "Good Intentions Are No Substitute for Good Practice," NSPE *Engineering Times*, Nov. 1981. A book review by the author of "Blaming Technology: The Irrational Search for Scapegoats," *St. Martin's Press*, 1981: c-169.

137. "Trend Analysis Program," Special Supplement, Spring 1976, *Institute of Life Insurance*, New York, NY: c-172.

138. Yoder, Edwin M., Jr., "Guest Comment," *Professional Engineer*, June 1979: c-78,173.

139. Pletta, Dan H., "Ethical Codes: Enforcement vs. Obedience," *Intersociety Conference on Engineering Ethics*, Baltimore, MD, May 1975, p. 82-83; Published completely as *ASME Preprint 75-TS-6* and partially as "Ethical Standards for the Engineering Profession: Where is the Clout?" in the *Professional Engineer*, July 1975: c-176,178,180.

140. Kemper, John D., "Teaching Professionalism and Ethics," ASCE *Civil Engineering*, Apr. 1978, p. 52-54: c-175.

141. Tofler, Alvin, "Learning for Tomorrow: The Role of the Future in Education," *Vintage Books*, Random House, New York, NY, 1974, p. 11: c-173.

142. ASCE *Report*, "Manual on Professional Civic Involvement," Task Committee on Professional Civic Involvement, John W. Frazier, Chm. 1974: c-177.

143. Binger, Wilson D., discussion of Arthus Casagrande's 2nd Terzaghi Lecture, ASCE *Journal of Soil Mechanics*, July 1965: c-181.

144. Videotape, "Ethics on Trial," produced by ASCE in 1978 depicts a reasonable facsimile of a case to illustrate the procedure its Committee on Professional Conduct follows when it recommends a hearing before the ASCE Board of

Direction for a member charged with violating its Code of Ethics: c-177.

145. Pletta, Dan H., "Professionalism Demands an Added Dimension," *Torch*, Winter 1980: c-183.

146. Bennis, Warren, "The Unconscious Conspiracy: Why Leaders Can't Lead," *American Management Association*, 1975: c-184.

147. Wenk, Edward, Jr., "Margins for Survival," *Pergamon Press*, New York, NY, 1979: c-185,186,188,210.

148. Mavis, F. T., "History of Engineering Education," *Journal of Engineering Education*, Vol. 43, 1952: c-93,185.

149. Nelson, Carl and Susan Peterson, "The Engineer as a Moral Agent," "Conflicts of Roles in Engineering Ethics," "A Moral Appraisal of Cost-Benefit Analysis," ASCE *Issues in Engineering*, Jan. 1982, p. 7-11: c-199.

150. Salk, Jonas, "Science, Technology and the Quality of Life," *Astronautics and Aeronautics*, Feb. 1970: c-199.

151. Marmor, Judd, "Psychiatry and the Survival of Man," *Saturday Review*, 22 May 1971: c-199.

152. Williams, L. Pearce, "Parallel Lives," *Executive*, Cornell University Graduate School of Business and Public Administration, Summer 1980: c-200,202.

153. Peter, Laurence J., "The Peter Principle," *W. Morrow Publ. Co.*, New York, NY, 1969: c-203.

154. Reports on "Profile of a Research and Development Executive --1978" and "Director--1981," *Heidrich & Struggles, Inc.*, New York, NY: c-190.

155. Greenberg, Daniel S., "Engineering Science in Decay: U.S. Technology Declining," *Engineering Times*, July 1980: c-188.

156. Tribus, Myron, "The Engineer and Public Policy Making," IEEE *Spectrum*, Apr. 1978: c-186,189.

157. Goshen, Charles E., "Engineering Characterology," *Journal of Engineering Education*, 1969: c-189.

158. Pascarella, Perry, "New Executives for New Times," *Industry Week*, Apr. 6, 1981: c-192.

159. Timberlake, Lloyd, "Poland's Pollution Crisis," *World Press Review*, Jan. 1982: c-193.

160. Sowell, Thomas, "Knowledge and Decisions," *Basic Books, Inc.*, New York, NY, 1980: c-193,205.

161. Rothbard, Murray, "Man, Economy and State," *Nash Publishing Co.*, Los Angeles, CA, 1961, Chapter 6: c-194.

162. Fink, Richard H., "Economic Growth: A Market Process Perspective," *Policy Report*, Cato Institute, Washington, D.C., Dec. 1981: c-194.

163. Watt, Simon, "Britain's Engineering: Shadow of the Past," *New Scientist*, July 9, 1981: c-212.

164. Williamson, Merritt A., "Engineering Management," *Encyclopedia of Management*, 3rd Edition: c-194.

165. Williamson, Merritt A., "Engineering Management in Responsible Charge vs. Administrative Management," NSPE *Proceedings* of the Conference and Workshop on Leadership and Pro-

fessionalism in Engineering Education, Nov. 1977: c-195.

166. Dougherty, Edward M., "Engineering Management in Industry," presented at the ASEM meeting, Nashville, TN, Sept. 1981: c-195,196.

167. News item, "Key to Productivity: U.S. Ignores Own Social Advances," quotes Robert Cole's comments at a congressional seminar on research and productivity, *Engineering Times*, Aug. 1980: c-197.

168. Clarke, Bruce, "The Challenge of Leadership," *Military Engineer*, July-Aug. 1961: c-203.

169. Simon, Julian L., "The Ultimate Resource," *Princeton University Press*, 1981: c-204.

170. Meadows, D. L. and D. H. Meadows, "Limits to Growth," *New American Library*, ISBM 0-451-09835-8, E 9835, New York, NY, 1972: c-204.

171. Vacca, Roberto, "The Coming Dark Age," *Doubleday & Company*, Garden City, NY, 1973: c-204.

172. Meadows, D. L., "The Only Game in Town--Social Aspects of the Sustainable State," *TechniUM*, College of Engineering, University of Michigan, Spring 1975: c-205.

173. "How to Make Slums and Create Barbarians," *Economic Education Bulletin*, American Institute for Economic Research, Great Barrington, MA, May 1981: c-206.

174. Unpublished letters from C. E. Hammett to the NSPE Education Committee, Apr. 12 & 14, 1975: c-206.

175. Pletta, Dan H., "Our Leaders are Led, Not Leading," *The Blacksburg* (VA) *Sun*, Mar. 18, 1979: c-208.

176. Maloney, Maureen," When the Buck Doesn't Stop Here," *Piedmont Magazine*, July 15, 1983, p. 34-37: c-23.

177. Ritter, Donald L., "Risk Analysis Research and Demonstration Act," *Congressional Hearing*, T174.5R55 Journal, YA Sci. 2:96/71, p. 113-118: c-19,23,94.

178. Newton, Lisa, "The Origin of Professionalism: Sociological Conclusions and Ethical Implications," *Business and Professional Ethics Journal*, Summer 1982, p. 33-43: c-16,17, 65.

179. Naisbitt, John, "Megatrends," *Warner Books*, New York, NY, 1982: c-17.

180. Inhaber, Herbert, "Energy Risk Assessment," *Gordon and Breach Science Publishers*, New York, NY, 1982: c-19,94.

181. Pletta, Dan H. and Charles H. Samson, "Engineering Unity: 20/20 Hindsight in the Year 2020," *Materials and Society*, Vol. 7, No. 2, May 1983, p. 201-215: c-44,45,73,74,96,172, 208.

182. Ogburn, William F. and Dorothy Thomas, "Are Inventions Inevitable? A Note on Social Evolution," *Political Science Quarterly*, Vol. 37, p. 83-98: c-55.

183. Blumenthal, Charles S., "Can't Repeal Laws of Physics: Politicians, Not Engineers, Cause Bridges to Fail," *Hartford* (Connecticut) *Courant*, July 17, 1983: c-32.

232

184. Lill, Richard A., "A Critical View of NSPE's Policy on Truck Weights," *Engineering Times*, May 1983: c-33.
185. Unger, Stephen H., "Controlling Technology: Ethics and the Responsible Engineer," *Holt, Rinehart and Wilson*, New York, NY, 1982: c-89,172.
186. Steffens, J., "Regulating to Protect the Public Interest," *Mechanical Engineering*, Nov. 1978, p. 85.
187. Hickerson, K. W., "Review Panel Urges Major Overhaul of Engineering Umbrella Group Activities," *Engineering Times*, Sept. 1983, p. 9: c-70.
188. Liberstein, Stanley H., "Who Owns What Is In Your Head: How to Protect Your Ideas," *Elsevier-Dutton Publ. Co.*, New York, NY, 1979: c-96.
189. Burr, Arthur A., "The Professional Program in Engineering at Rensselaer," Challenge for the Future: NSPE Professional School Task Force *Report*, 1976: c-135.

Appendix A: c-67-68
      B: c-88,165,174,
      C: c-175
      D: c-175
      E: c-175

Index

236

238

239

medical education 66
Mensch, Gerhard 55
mergers 191,192
Mianus River Bridge failure 32
Michelangelo 205
Military Academy, U.S. 58
minimum wage law 206
Mintz, Morton 97
Mises, Ludwig von 23,28
missions
  corporate 184
  societal 184,198
  technological 1
mobility of engineering gradu-
  ates 146
Mohammed 203
Monastaries
  as Medieval schools 122
monopolies 47,49
moral issues 161
  standards 162
Morrill Land Grant Act 125
Mozart 205
Nader, Ralph 53,59,87,89,172,
  176,180,186
Naisbitt, John 17,65
Nash, Ogden 15
National Academy of Engineers
  142
  1969 Proceedings on Leader-
    ship 37
National Commission on Product
  Safety 115
National Science Fdn. Report
  152
NCEE 68,69,179,194
Nelson, Carl 198
Newcomen's steam engine 48
Newton, Issac 57,201
Newton, Lisa 16,17,65
Norris, William C. 191
Norwich University 158
NSPE
  as a member of AAES 70
  Board of Ethical Review 178
  ethical violations 178
  liaison agreements 72
  long range strategy 128,131,
    133
  professional schools 128,
    131,133

professional unity 72
Public Advocacy Review Board
  88
qualifications for practice
  6,16,39,41,83-85,94
Supreme Court decision 168
objectives
  flexible 200
  of corporations 190
  of engineering education 117
  of management 187
  of professional schools 138
obligation, professional--see
  professional
obsolescence 23
Occupational Safety and
  Welfare
  President's Report 115
Office of Technology Assess-
  ment 11,83,143
Ogburn, W. F. 55-56
ombudsman 4,13,59,87-90
  functions 87,164
  professional boards of 88
Oparin, A. I. 103
Order of the Engineer 175
  Obligation of Appendix C
Oxford University 123
Panama Canal 209
Page, Bruce 59
Palestine--early education 120
Paris, University of 122
participatory membership 78
Pascarella, Perry 192
Perronet, Jean R., founder of
  Ecole 124
personal integrity 175
Pertschuk, Michael 172
Peter (Lawrence J.) Principle
  203
Peterson, Susan 198
Petkas, Peter 59
Piehler, R. H. 115
Plato 120
Poland, environmental pollu-
  tion 193
political action committees
  206
polls, of engineers to formu-
  late technological
  policies 78,206

240

241

Veblen, Thorstein 23,62
Vinci, Leonardo da 52
vocation 39
Walker, W. A. 130
Wallich, H. C. 36
Warren, Earl 162
Washington, George 200-02
Washington, University of;
    WISE Fellowships 211
Watt, Simon 212
Watt's steam engine 48
Weinstein, A. S. 115
welfare, public 51
Wenk, Edward, Jr. 185,186,188,
    210
whistle blowers 36,176,180
White House Fellows 143,211
Whitehead, Alfred North 150
Whitelaw, Robert L. 13,37,60
Whitney, Eli 58
Whittle, Sir Frank 97
Wickenden, William E. 87,130
Wilkie, Wendell L. 105
Williams, L. Pearce 200-02
Williamson, Merritt A. 194,195
WISE Fellowships 211
Wisely, W. H. 6,77,78
Wright, Patrick J. 62
Xenophone 122
Yoder, Edwin M., Jr. 78,172-73
zeal, professional--see pro-
    fessional
Zuesse, Eric 89

243

About the Author

Dan H. Pletta has taught at four universities and at the U.S. Military Academy. He was head of Engineering Science and Mechanics for 22 years at Virginia Polytechnic Institute and State University. There, in 1970, he became one of the original five University Distinguished Professors appointed and is now Emeritus. He earned B.S. and C.E. degrees from Illinois, and an M.S. from Wisconsin. Besides teaching, his experience includes design and construction in civil engineering, consulting as a stress analyst, and World War II service as an Ordnance Officer.

Recent awards include an Engineering News Record Citation for Service to the construction industry, Engineer-of-the-Year Award of the Virginia Society of Professional Engineers, and the Alumni Honor Award for Distinguished Service from the College of Engineering at the University of Illinois. He has served on the National Boards of Direction of the National Society of Professional Engineers (NSPE) and of the American Society of Civil Engineers (ASCE), as an NSPE Vice President, as Chairman of the ASCE Committee of Professional Conduct, and on many committees of these and other societies.

He is a registered professional engineer and the author of numerous technical articles, an engineering handbook chapter and two texts on engineering mechanics, as well as papers on professionalism and ethics. In 1979 he was elevated to Honorary Membership in ASCE.